PROGRESS IN COMPACT ANTENNAS

Edited by **Laure Huitema**

Progress in Compact Antennas
http://dx.doi.org/10.5772/56980
Edited by Laure Huitema

Contributors

Nacer Chahat, Alexander Krasnok, Pavel Belov, Andrey Miroshnichenko, Arseniy I. Kuznetsov, Boris Lukyanchuk, Yuri Kivshar, Laure Huitema, Henrique Manuel Salgado, Qi Luo, Jose Pereira, Fabio Viola, Pietro Romano, Mohamed Himdi

Published by InTech

Janeza Trdine 9, 51000 Rijeka, Croatia

Publishing Process Manager Iva Lipovic
Technical Editor InTech DTP team
Cover InTech Design team

First published September, 2014
Printed in Croatia

Additional hard copies can be obtained from orders@intechopen.com

Progress in Compact Antennas, Edited by Laure Huitema
p. cm.
ISBN 978-953-51-1723-0

For EU product safety concerns:
IN TECH d.o.o., Prolaz Marije Krucifikse Kozulić 3, 51000 Rijeka, Croatia,
info@intechopen.com or visit our website at intechopen.com.

Meet the editor

Laure Huitema was born in Limoges, France, in 1984. She received the M.S and Ph.D. degrees in telecommunications high frequencies and optics from Limoges University in 2008 and 2011 respectively.
From 2011 to 2012, she was a Postdoctoral Research Fellow at the Atomic Energy Commission (CEA), Laboratory of Electronics and Information Technology (LETI), Grenoble, France. She is currently an associate professor in the OSA (Wireless communications and Effect of EM Wave) Department of XLIM Research Institute. Her research interests include active antennas, Dielectric Resonator Antennas and also multiband antennas. As a result of her work, she was the recipient of the best student paper award at the 2010 IEEE International Workshop on Antenna Technology (iWAT 2010) and of the best student paper award at the 2010 Journées de Charactérisation Microondes et Matériaux (JCMM 2010).

Contents

Preface

Compact antennas are a subject of growing interest from industry and scientific community to equip wireless communicating objects. Although a lot of effort has been made in the design of mobile terminal antennas, major advances are needed for emerging applications such as pervasive wireless sensor networks. The need for high performance small antennas and RF front ends is the challenge for future and next generation mobile devices.

In addition to being compact, major keys for antennas designers are the implementation of multi-functionality and broadband properties.

This book brings together the rapidly growing body of knowledge on compact antennas into a single comprehensive volume. The Progress in Compact Antennas book is designed to meet the needs of electrical engineering and physics students at the senior undergraduate and beginning graduate levels, and those of practicing engineers.

Information contained in this book is the result of ongoing research investigations on compact antennas. I would like to acknowledge the work of all authors who have contributed to the realization of this book.

I am also grateful to the staff of InTech, especially Ms. Iva Lipović, for her interest, support, cooperation, and production.

Laure Huitema
Xlim Research Institute,
University of Limoges,
France

Chapter 1

Compact Antennas — An overview

L. Huitema and T. Monediere

Additional information is available at the end of the chapter

http://dx.doi.org/10.5772/58837

1. Introduction

Antenna size reduction is restricted by fundamental physical limits [1-3], in terms of trade-off between radiation performances and impedance bandwidth. Miniaturization of devices leads to the reduction of antennas which becomes one of the most important challenges [4]. Limitations in terms of bandwidth and efficiency suggest an analysis with respect to fundamental limits [5]. Although interests are often focused on the impedance bandwidth, many studies deal with the radiation quality factor Q. Some papers [6] have been concluded that the impedance bandwidth BW equals 1/Q. The minimum Q value reachable by an infinitesimal electric dipole, or similarly by the azimuthally symmetric TM_{10} spherical mode, has been investigated thoroughly.

Hansen and Best [7] have shown that the lower bound on Q, deriving from Chu's analysis, is depending on the expense of efficiency as shown by the equation:

$$Q_{lb}=\eta\left(\frac{1}{(ka)^3}+\frac{1}{ka}\right)$$

where a is the minimum radius of the sphere enclosing the antenna and k is the wave number ($k=2\pi/\lambda$).

The Figure 1 shows that it is very difficult to have a wide bandwidth (low Q-factor), while reaching a good efficiency for miniature antennas (k.a around 0.2). Thus, the miniaturization of antennas implies them to suffer of both limited efficiency and low bandwidth.

Since many years the scientific literature addresses some approaches concerning miniaturization techniques. The goal is to decrease the electrical size of the radiating element. This chapter will draw up a survey of compact antennas in practical settings and the most common miniaturization techniques listed below:

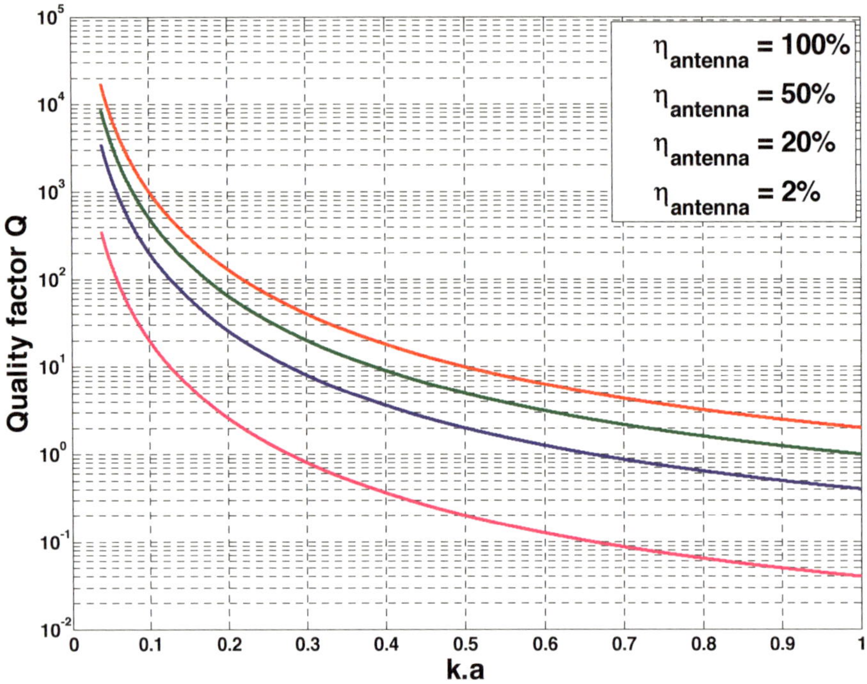

Figure 1. Quality factor according to the antenna dimensions and efficiencies

- Folded configurations [8-10]
- Surface etching techniques [11-14]
- Shorting walls or pins [15-16]
- The use of high dielectric constant materials or magneto-dielectric materials [17-22].
- Loading the radiating element with active components [23-26].
- Creation of hybrid modes with particular boundary conditions in dielectric resonator antennas. It allows choosing their resonance frequencies (for multiband or wide impedance bandwidth) [27].

We will start this chapter by detailing wire antennas. Indeed, after explaining the classical dipole antenna, we will show how to miniaturize this kind of antennas based on shape design such as bending, folding and meandering. The second part will detail planar antennas. We will see the impact of materials properties under the patch antenna hat, i.e. dielectric or magneto-dielectric materials. Then, planar miniature antennas will be shown, e.g. Planar Inverted F Antenna (PIFA) and monopolar wirepatch antenna. The third part will exhibit Dielectric Resonator Antennas and how to use this kind of antennas for low frequency band application while having compact sizes. Finally, the last part will summary all the antennas presented in this chapter, while showing their main settings.

http://dx.doi.org/10.5772/58837

2. Wire antennas — Miniaturization techniques

2.1. Classical wire antenna: The dipole antenna

The dipole antenna has been developed by Heinrich Rudolph Hertz around 1886 and still remains the most widely used antenna (Figure 2). It owns two identical (same length) and symmetrical metal wires, and its feeding device is connected at the center of the dipole, i.e. connected to the two adjacent wires ends. The dipole working results of a standing wave phenomenon depending on its length. The antenna fundamental mode occurs when the whole antenna is a half-wavelength long.

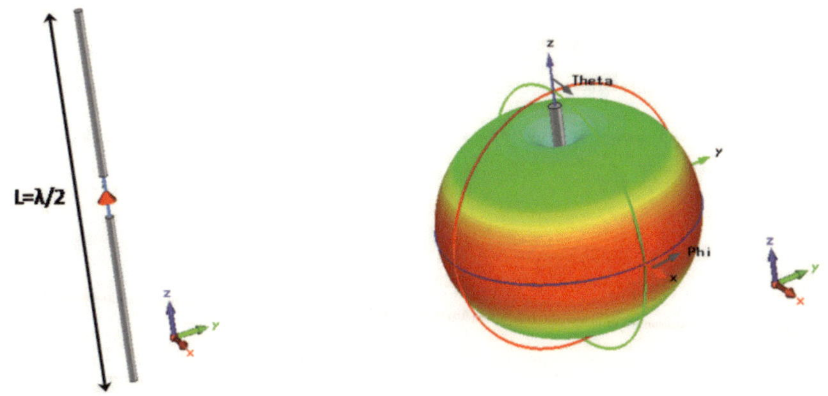

(a) (b)

Figure 2. Dipole antenna shape (a) and its 3D radiation shape (b)

The radiated field of the dipole antenna working on its fundamental mode has a linear polarization. As shown in Figure 2, its radiation pattern is maximum at right angles to the dipole and drops off to zero on the dipole's axis. Its maximum directivity equals 2.15dBi. The impedance bandwidth of this kind of antenna is quite wide since it is between 10% and 20% (it depends on the wire's radius) [28].

2.2. The monopole antenna

By adding a perpendicular ground plane at the center of the dipole antenna, its length can be divided by two: that is the monopole antenna. Theoretically, this ground plane is considered as an infinite Perfect Electric Conductor (PEC) plane. In this case, the current in the reflected image [29-30] has the same direction and phase as the current in the dipole antenna. Thus the quarter-wavelength monopole and its image together form a half-wavelength dipole that radiates only in the upper half of space (see Figure 3).

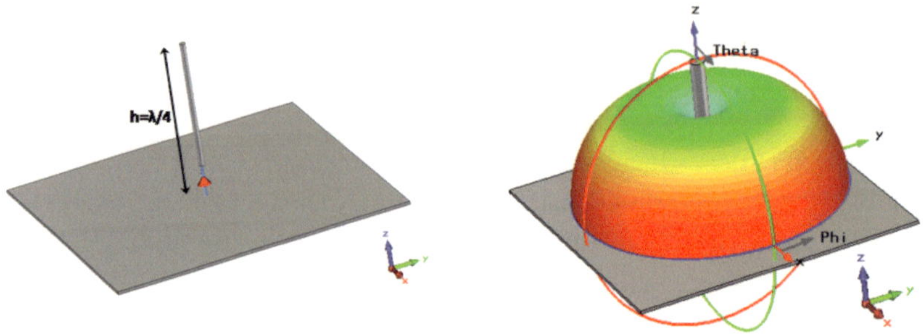

(a) (b)

Figure 3. Monopole antenna shape (a) and its 3D radiation shape (b)

Therefore it presents a 3 dB gain higher than the dipole antenna. The radiation resistance is proportional to $(h / \lambda)^2$, this latter is therefore decreasing the square of the antenna height h.

The first mobile phones were using this kind of antennas to receive the Global System for Mobile Communications (GSM) (see Figure 4).

Figure 4. Monopole antenna integrated inside a mobile phone

In practice, the finite ground plane disturbs the radiation pattern and the maximum directivity is decreasing. The monopole antenna bandwidth is quite the same as the dipole, i.e. up to 20%.

2.3. Inverted L and F antennas (ILA and IFA)

To reduce the monopole antenna global dimensions, we can bend the wire to be parallel to the ground: that is the Inverted L Antenna (ILA) [31]. Its design is depicted Figure 5, there is both a vertical and a horizontal parts. Since its electrical length is the same than the monopole, its resonance frequency is also the same. The radiation resistance is proportional to $(h / \lambda)^2$, with h the length of the vertical part (see Figure 5). Actually, the horizontal part occurs as a capacitive charge and this makes the antenna difficult to match on 50Ω. Therefore, the antenna bandwidth is very low and does not exceed 1% [31-33].

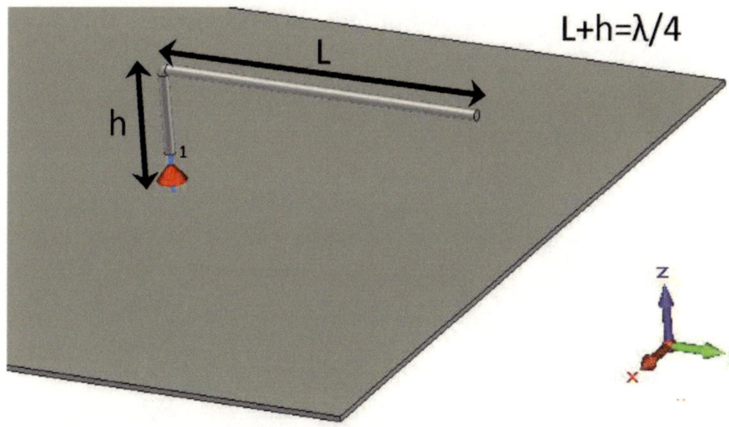

Figure 5. Inverted L Antenna (ILA) shape

Adding a ground wire on the horizontal part facilitates the ILA matching. This new antenna design is called Inverted F Antenna (IFA) (Figure 6). This wire is equivalent to a self-inductance in parallel with the capacitance of the horizontal wire. That involves a parallel resonance at low frequencies.

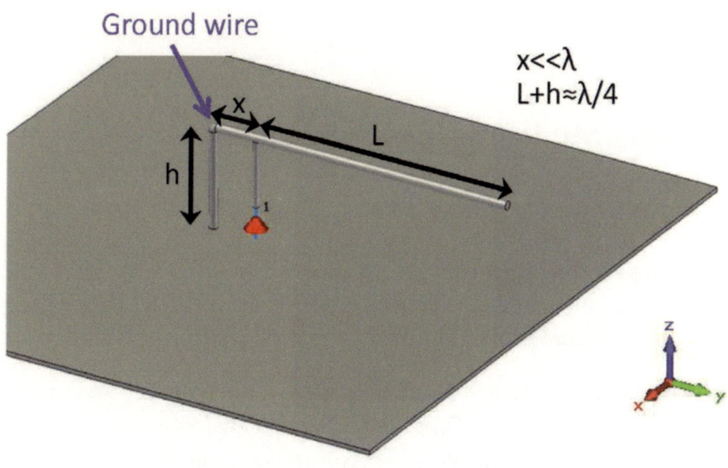

Figure 6. Inverted F Antenna (IFA) shape

To fit the input impedance around 50Ω, we can adjust wire's parameters (radius, distance with the feeding point,...). However, typical impedance bandwidths are around 2% or 3% [34-35].

2.4. The helical antenna

This kind of antenna allows reducing the physical length of an antenna (Figure 7). Basically, its fundamental mode is due to a quarter wavelength resonance. However, its bending structure can involve some capacitive and/or inductive resonances. This antenna has been widely used in mobile phone devices.

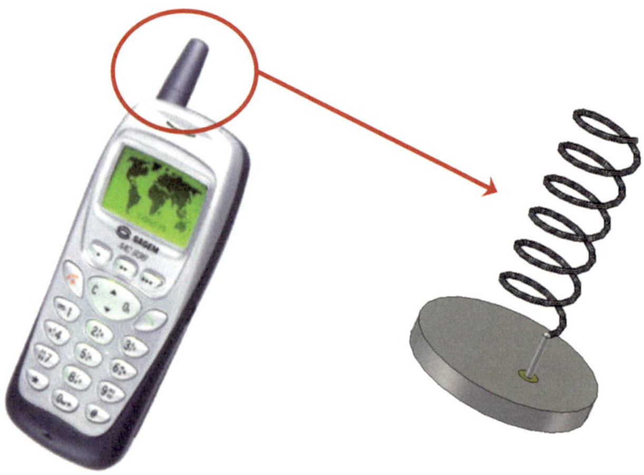

Figure 7. Helical antenna

Global dimensions of this kind of antennas are around $\lambda_0/10$ for its height with a $\lambda_0/40$ diameter. Its typical impedance bandwidth is up to 8%. Thus this antenna is covering the entire GSM band [36-37].

3. Planar antennas – Miniaturization techniques

3.1. Classical planar antenna: the patch antenna

The patch antenna was introduced by John Q. Howell in 1972 [38]. This kind of antenna presents a metallic top hat mounted on a dielectric substrate. Its lower face is the ground plane and its feeding can be a coaxial probe (Figure 8), a microstrip line or a coplanar waveguide.

The two metal sheets together form a resonant part of a microstrip transmission line with a length equals to a half of wavelength. Thus, its higher dimension is equal to $\lambda_g/2$, with λ_g the the guided wavelength. A simple patch antenna radiates a linearly polarized wave and its radiation can be regarded as a result of the current flowing on the patch and the ground plane. Thus its maximum gain is relative to the vertical axis of the patch and can reach 7 or 8 dB. The

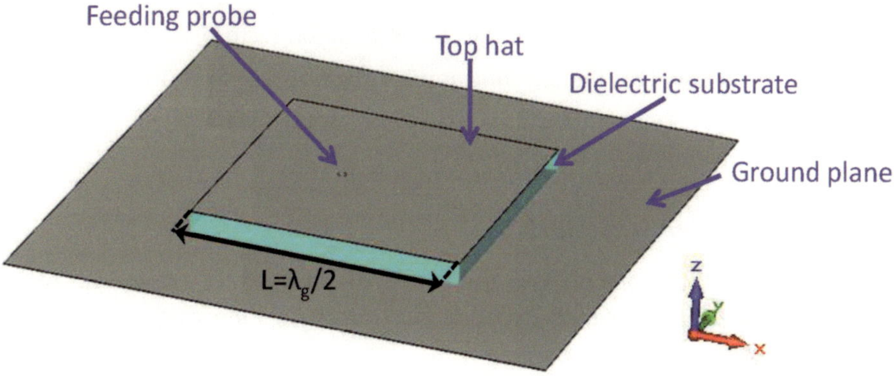

Figure 8. Patch antenna

impedance bandwidth is between 1% and 4% and depends on both the dielectric permittivity and the thickness of the substrate.

3.2. Miniaturization techniques of planar antennas

3.2.1. Use of materials

Using a material allow the reduction of the guided wavelength and thus the physical length of an antenna. Indeed, the patch antenna length is directly proportional to the refractive index of the dielectric substrate $n=\sqrt{\varepsilon_r.\mu_r}$. Using a dielectric material is the most common method to reduce the antenna size [39-40]. In this sub-section, we will show on one hand the antenna size reduction for a planar antenna printed on a dielectric substrate and on the other hand on a magneto-dielectric material substrate.

- **Using a dielectric material**

We consider the patch antenna presented Figure 8 with a 4cm-length and a 3mm-thickness. To show the impact of a dielectric substrate we consider two different cases:

- Substrate with low dielectric permittivity $\varepsilon_r=1$.
- Substrate with higher dielectric permittivity $\varepsilon_r=9$.

The Table 1 summarizes main results for patch antennas with strictly same dimensions.

	Resonance frequency	Matching frequency	Impedance bandwidth
Patch antenna with $\varepsilon_r=1$	3.7 GHz	3.8 GHz	4.1 %
Patch antenna with $\varepsilon_r=9$	1.3 GHz	1.32 GHz	0.98 %

Table 1. Comparison between two antenna patches with same dimensions printed on different substrates

As we can see in this table, the resonance frequency is divided by three with the increase of the dielectric permittivity value. Indeed, the miniaturization factor is close to the refractive index $\sqrt{\varepsilon_r.\mu_r}=\sqrt{9}=3$. However, the antenna miniaturization involves the reduction of its performances as its impedance bandwidth is divided by four.

Another solution is to use a magneto-dielectric material.

- **Using a magneto-dielectric material**

Hansen and Burke [41] have expressed the zero-order impedance bandwidth of a patch antenna printed on a t-thick magneto-dielectric material by the following equation:

$$BW = 96\sqrt{\frac{\mu}{\varepsilon}}.\frac{t}{\lambda_0}\Big/\sqrt{2}\left(4+17\sqrt{\varepsilon.\mu}\right) \tag{1}$$

Thus, compared to high dielectric permittivity, high permeability materials allow to reduce the size of a patch antenna without decreasing its relative impedance bandwidth. In [42], Niamien et al. investigates magneto-dielectric materials losses and provides expressions of antenna impedance bandwidth and efficiency according to both dielectric and magnetic losses for a patch antenna. They showed that both the radiation efficiency and the impedance bandwidth increase with the permeability.

Considering the previous patch antenna (with a 4cm-length and 3mm-thick) by changing the dielectric material by a magneto-dielectric material, we obtain the Table 2 results.

	Resonance frequency	Matching frequency	Impedance bandwidth
Patch antenna with $\varepsilon_r=9$ and $\mu_r=1$	1.3 GHz	1.32 GHz	0.98 %
Patch antenna with $\varepsilon_r=4$ and $\mu_r=2.25$	1.35 GHz	1.38 GHz	1.87 %
Patch antenna with $\varepsilon_r=3$ and $\mu_r=3$	1.37 GHz	1.41 GHz	2.82 %
Patch antenna with $\varepsilon_r=2.25$ and $\mu_r=4$	1.38 GHz	1.45 GHz	3.29 %
Patch antenna with $\varepsilon_r=1$ and $\mu_r=9$	1.31 GHz	1.65 GHz	4.66 %

Table 2. Comparison between antenna patches with same dimensions printed on different substrates

This table compares patch antenna results with a same refractive index $n=\sqrt{\varepsilon_r.\mu_r}=3$. It should be noticed that all the materials are considered without any loss.

Therefore the comparison between the dielectric and magneto-dielectric materials shows that using the latter in a patch antenna allows increasing its impedance bandwidth. A patch antenna printed on a magneto-dielectric material presents the same miniaturization factor and allows the increase of its impedance bandwidth.

3.2.2. *Modification of the antenna shape*

- **Notches integration**

The integration of notches on the antenna top hat is often used. It allows to artificially increase the electrical length of the radiating element by extending the current "path" on this element (Figure 9).

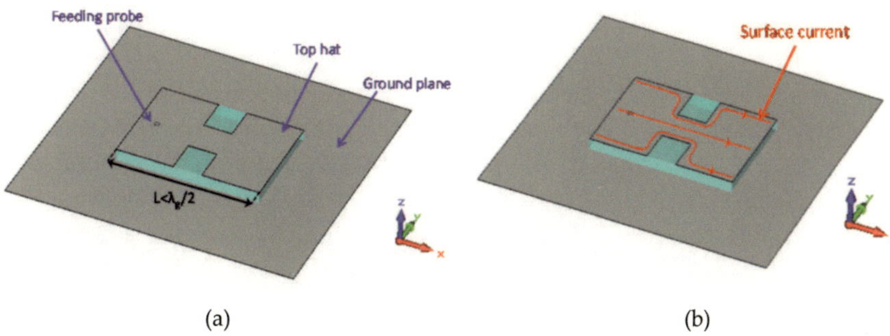

Figure 9. Integration of notches on antenna's top hat (a) Surface current lines (b)

The radiating element dimensions can be reduced up to 50% comparing with a classical patch antenna.

- **Meander antenna**

As for the helical antenna, the meander antenna allows decreasing the physical length of a planar antenna. The advantage is that this antenna is planar and thus easy to integrate inside a mobile phone. We can present on the Figure 10 a widely used meander antenna for the GSM reception on mobile phones. It is printed on a 0.8mm-thick FR4 substrate and is matched on 1%-bandwidth around 900 MHz with $\lambda_0/3$ x $\lambda_0/5$ dimensions.

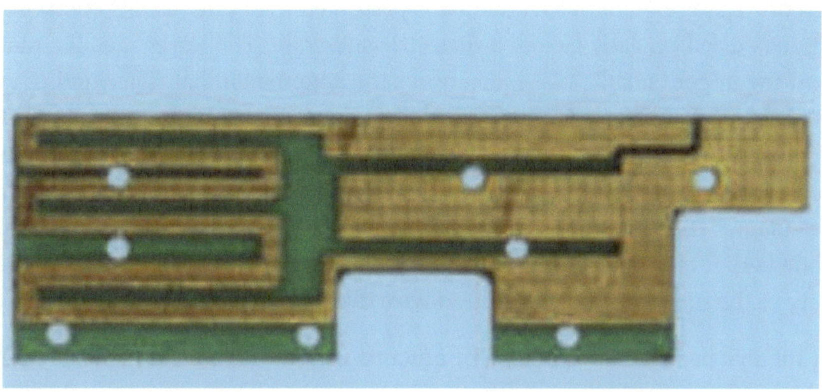

Figure 10. Meander antenna integrated within a mobile phone for the GSM reception

3.2.3. Short cut insertion

- **Planar Inverted F Antenna**

As for the dipole and the monopole, it is possible to integrate a metallic plate inside the patch antenna in order to divide its main dimension by two. Indeed, on the fundamental mode of the patch, we can integrate a short cut where the electric filed is null. To manage to match the antenna the metallic plate dimensions have to be optimised (Figure 11).

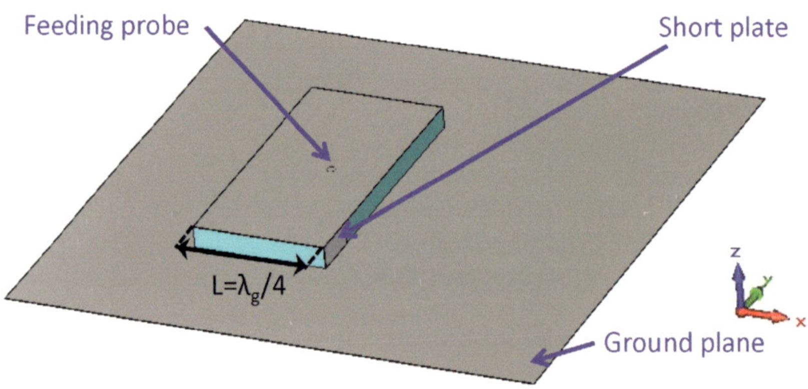

Figure 11. Planar Inverted F Antenna (PIFA)

In [43], a Planar Inverted Antenna with $\lambda_0/6$ x $\lambda_0/8$ x $\lambda_0/35$ dimensions at 2.45GHz is matching over a 6.5% impedance bandwidth.

- **Monopolar wire patch antenna**

The design of a classical wire patch antenna is presented in Figure 12. It is composed by two metallizations etched on each face of a dielectric substrate. The lower metallic plate acts as ground and the upper metallic plate constitutes the antenna top hat. This kind of antenna is fed by a coaxial probe which is connected to the top hat through the ground plane and the dielectric substrate. The ground wire acts as a short-circuit to the capacitance of the antenna constituted by the top hat above the ground plane and allows achieving a new low-frequency parallel resonance. The resonance frequency is smaller than the classical antenna fundamental cavity mode [44]. It is primarily set by the size of the top hat, the height of the antenna, the permittivity of the substrate and the ground wire diameter.

The main antenna parameters to adjust the antenna impedance matching to 50Ω are:

- The ground wire radius. The smaller the radius is, the higher the maximum of the input impedance real part is.

- The radius of the feeding probe. The higher the radius is, the lower the input impedance imaginary part is.
- The ground wire – feeding probe separation. The Q-factor is increasing when the length between the ground wire and the feeding probe core is increasing.

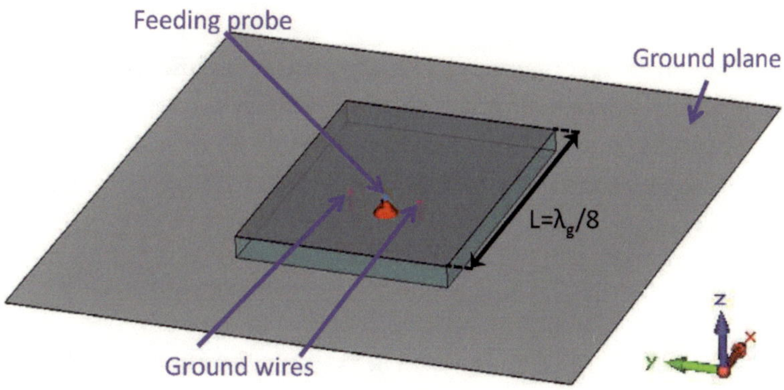

Figure 12. Monopolar wire patch antenna

As presented in [45-46], the use of a closed slot into the antenna top hat involves a significant reduction of the resonant frequency (Figure 13). Indeed, the introduction of a slot in the hat of the antenna changes the equivalent capacitance of the antenna short-circuited hat by increasing its value.

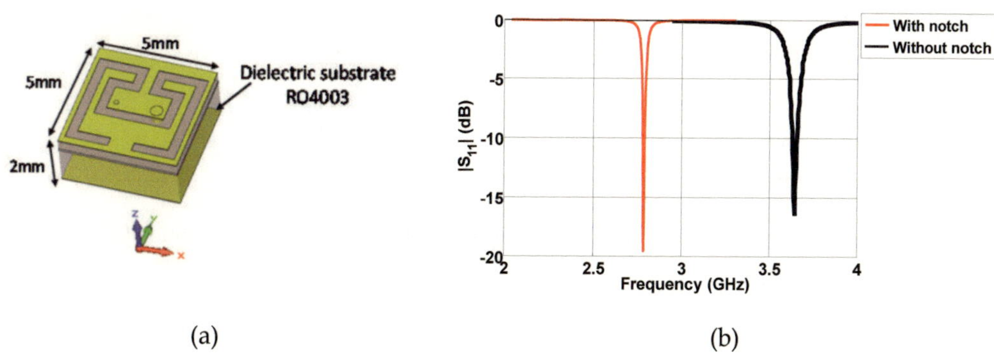

Figure 13. Monopolar wire patch antenna with a notch (a) and corresponding $|S_{11}|$ parameters

The longer the electrical length of the slot is, the lower the resonant frequency is. We can compare the $|S_{11}|$ parameters of the wire patch antenna presented in Figure 13 with and without the notch [47]. As expected, adding a notch inside the wire patch antenna top hat allows decreasing the working frequency but also the impedance bandwidth.

4. Dielectric resonator antennas (DRAs) – Miniaturization techniques

The design of a DRA in any geometry must satisfy various specifications including: the resonant frequency, the impedance bandwidth, the field distribution inside the resonator and also the radiated field. The intent of this part is to provide an overview of main findings of investigations on simple-shaped DRAs. Then, it will deal with the different miniaturization techniques of DRAs.

4.1. DRAs characteristics

A non-exhaustive list of main simple-shaped DRAs characteristics is described below:

- The main dimension of a DRA is proportional to $\lambda_0 / \sqrt{\varepsilon_r . \mu_r}$ where λ_0 is the free-space wavelength at the resonant frequency, ε_r and μ_r are respectively the dielectric and the magnetic constant of the material. In a dielectric material case, $\mu_r = 1$ and the main dimension of a DRA is proportional to $\lambda_0 / \sqrt{\varepsilon_r}$.
- The radiation efficiency of the DRA is highly depending on the material losses. In case of a low-loss dielectric material, DRAs allow to achieve better efficiency than other kind of antennas because of minimal conductor losses associated with a DRA.
- For a given dielectric constant, both resonant frequency and radiated Q-factor are defined according to the resonator dimensions. That allows having a great flexibility and some degrees of freedom to design such an antenna.
- Another degree of freedom is the large spectrum of available dielectric materials. That allows doing the best trade-off between dimensions and impedance bandwidth according to the intended application.
- A number of modes can be excited within the DRA, many of them provide dipolar-like radiation characteristics.
- The most common targeted frequencies presented by the research literatures are ranging from 1GHz to 40 GHz.
- For a given DRA geometry, the radiation patterns can be made to change by exciting different resonant modes.

A large number of DRA excitations are currently used, e.g. microstrip line, coaxial probe excitation, coplanar waveguide… The next subsection will deal with the most commonly used excitations.

4.2. DRAs miniaturization techniques

This subsection examines techniques to design compact DRAs. Targeted applications are mobile handsets or wireless tablet. There are several techniques to make DRAs more compact. By adding metal plates, inserting a high permittivity layer (multisegment DRA) or removing portions of the DRA, a significant size reduction can be achieved.

- **Addition of a metallic plate on a DRA face**

The rectangular DRA shape has been studied in the first part. The perfect metallic wall implies that electric fields are normal to this conductor, while magnetic fields are tangential. E and H fields presented Figure 14 assume that a metallic plate can be inserted in the middle of the DRA according to the y-component. The principle is detailed and explained by the Figure 14. It also shows the E and H fields of the TE_{111} mode.

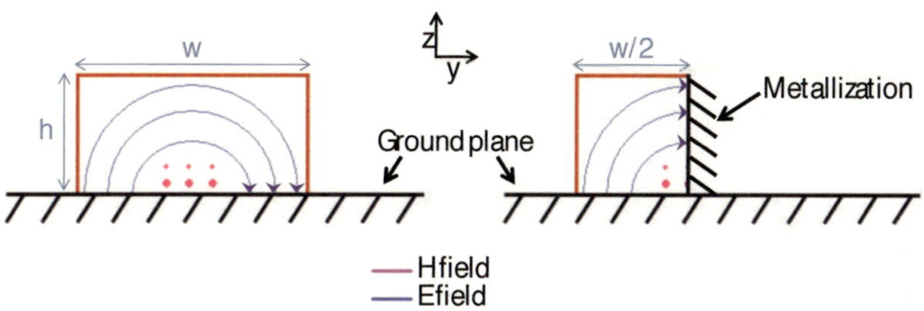

Figure 14. Integration of a metallic plate

By applying the image theory, it is possible to insert a metal plate in the y=w/2 plane. The Table 3 extracted from [48] shows the influence of the metallic plate insertion on resonant frequency and impedance bandwidth.

ε_r	w (cm)	d (cm)	h (cm)	Metallization	f_0(GHz)	Bandwidth
12	2.75	2.75	2.95	No	1.98	10%
12	2.75	2.75	2.95	Yes	1.24	5.6%

Table 3. Influence of the metallic plate insertion on both resonant frequency and impedance bandwidth

Thus, the metal plate insertion allows dividing by two the DRA size, while reducing the resonant frequency. However, as pointed by the Table 3, the metallic plate insertion involves also the decrease of the impedance bandwidth.

- **Multisegment DRA**

Another way to decrease the DRA size is to insert different substrate layers as illustrated Figure 15.

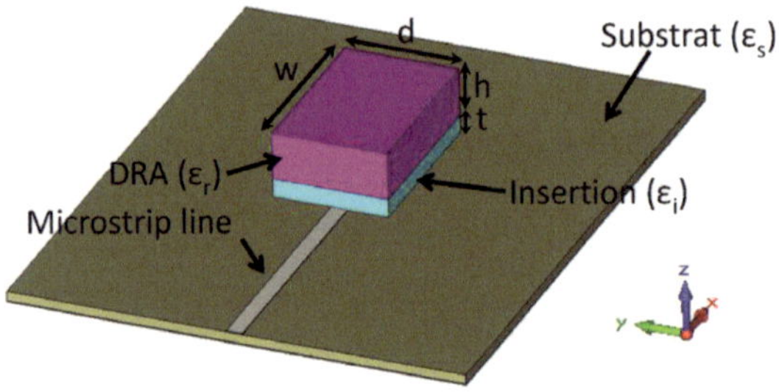

Figure 15. Multisegment DRA

It allows achieving strong coupling when the first insertion has a relatively high dielectric permittivity. This technique is detailed in [48] and [49]. The Table 4 summarizes a parametrical study done in [49] for one layer inserted (Figure 15) with w=7.875 mm, d=2 mm, h=3.175 and ε_r=10. It is mounted on a 0.762 mm height substrate of permittivity ε_s=3. The TE_{111} mode of the DRA is excited with a 50Ω microstrip line.

t (mm)	ε_i	Measured f_0(GHz)	Bandwidth
0	-	15.2	21%
0.25	20	14.7	18%
0.635	20	14.5	18%
1	20	13.9	16%
0.25	40	14.7	20%
0.635	40	13.7	13%
1	40	12.9	5%
0.25	100	14.7	16%
0.635	100	13.1	7%
1	100	10.8	5%

Table 4. A parametrical study done in [31] for one layer inserted

Thus, a thin layer insertion allows improving the coupling of modes inside the DRA while decreasing the resonant frequency thanks to the decrease of the effective dielectric permittivity of the DRA. As the previous technique, the downside is the decrease of the impedance bandwidth.

http://dx.doi.org/10.5772/58837

5. Summary of compact antennas performances

In this part, as a first conclusion we can summary the main performances of the previous presented antennas.

	Substrate	Dimensions	Impedance bandwidth	Radiation	Directivity (dBi)	Total efficiency
Dipole	Air	$\lambda_0/2$	10% à 20%	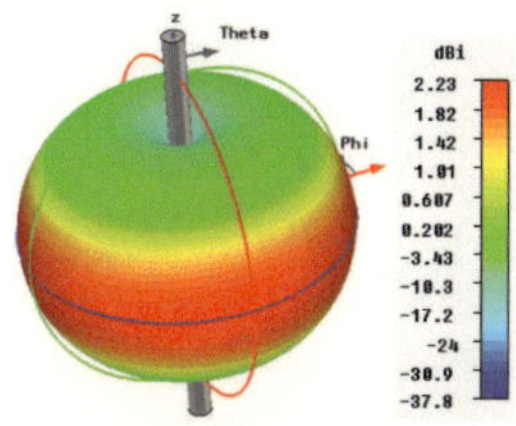	2.1	99%
Monopole	Air	$\lambda_0/4$	10% à 20%	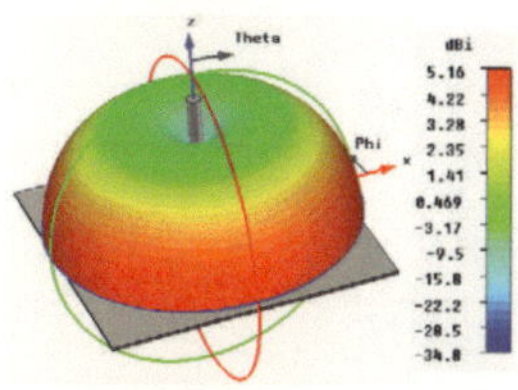	5.1	99%
ILA	Air	$\lambda_0/20 \times \lambda_0/4$	1%	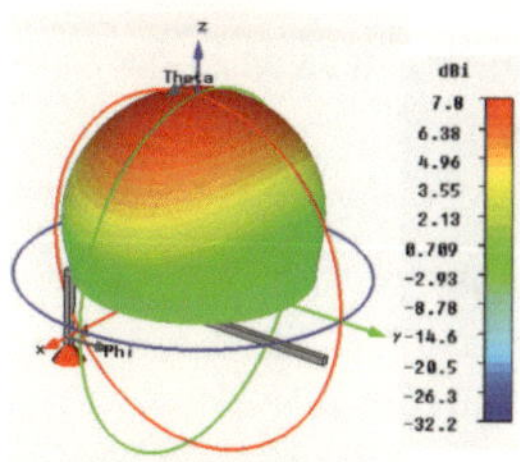	7	98%
IFA	Air	$\lambda_0/20 \times \lambda_0/4$	2%	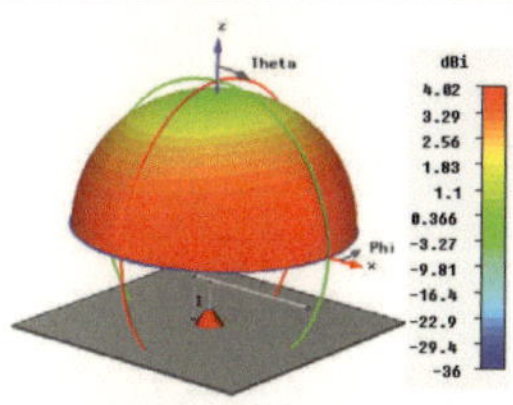	4	98%
Helical	Air	$\lambda_0/10 \times \lambda_0/40$	7%	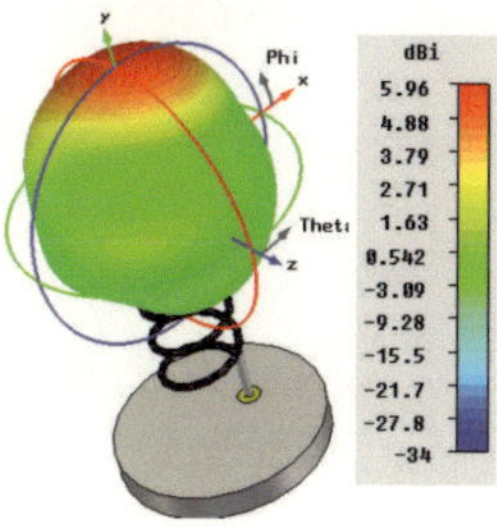	6	98%

	Substrate	Dimensions	Impedance bandwidth	Radiation	Directivity (dBi)	Total efficiency
Patch	Air	$\lambda_0/2 \times \lambda_0/2 \times \lambda_0/27$	4.1%	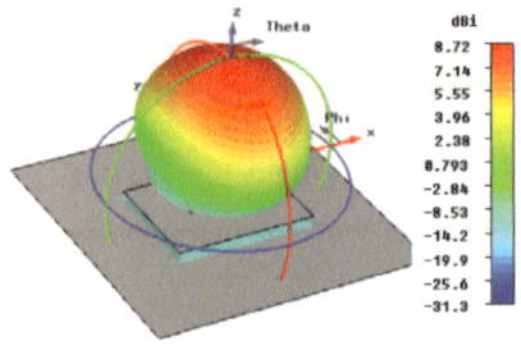	8.8	98%
Patch	$\varepsilon_r=9$ ($\tan\delta=5.10^{-4}$)	$\lambda_0/6 \times \lambda_0/6 \times \lambda_0/76$	0.98%	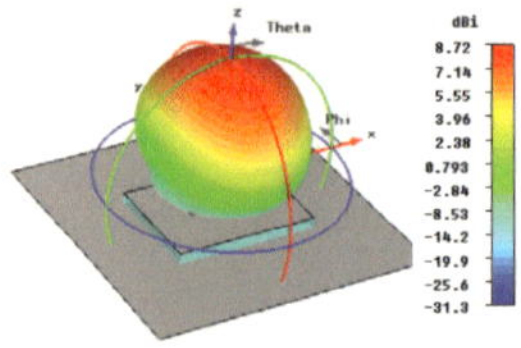	7	89%
Patch	$\varepsilon_r=2.25$ - $\mu_r=4$ ($\tan\delta_\varepsilon=5.10^{-4}$ $\tan\delta_\mu=5.10^{-4}$)	$\lambda_0/6 \times \lambda_0/6 \times \lambda_0/76$	3.29%	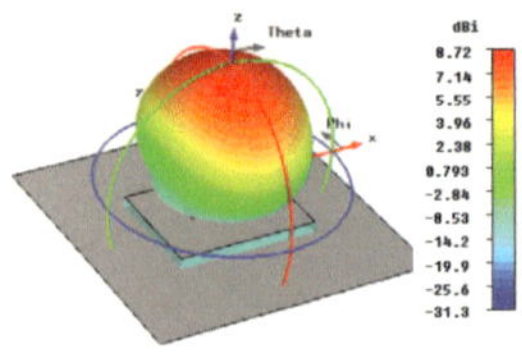	7.2	92%
Patch	$\varepsilon_r=1$ $\mu_r=9$ ($\tan\delta_\varepsilon=5.10^{-4}$ $\tan\delta_\mu=5.10^{-4}$)	$\lambda_0/6 \times \lambda_0/6 \times \lambda_0/76$	4.66%	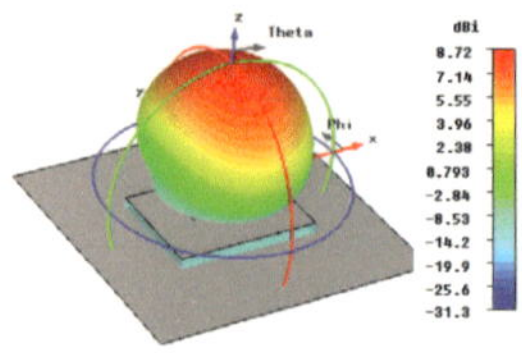	7.5	95%
Patch with notches	Air	$\lambda_0/4 \times \lambda_0/5 \times \lambda_0/68$	1.2%	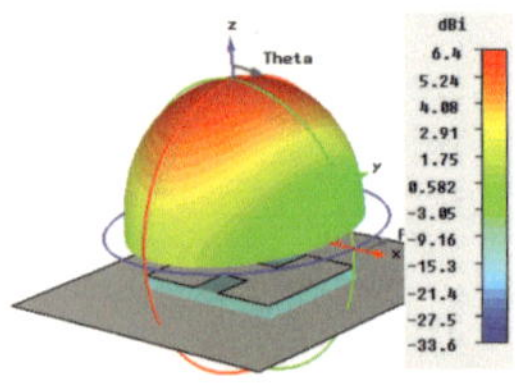	6.4	95%
PIFA	Air	$\lambda_0/6 \times \lambda_0/8 \times \lambda_0/35$	6.5%	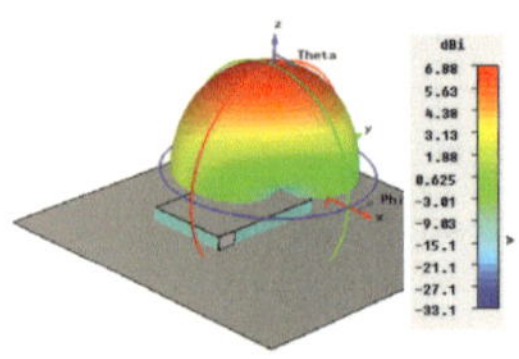	6.9	97%
Wirepatch antenna	Air	$\lambda_0/8 \times \lambda_0/8 \times \lambda_0/17$	3%	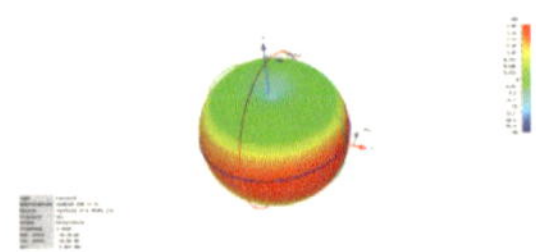	4	97%

	Substrate	Dimensions	Impedance bandwidth	Radiation	Directivity (dBi)	Total efficiency
DRA (2 layers)	ε_{layer1}=100 ε_{layer2}=10 ($\tan\delta_\varepsilon$=5.10^{-4})	$\lambda_0/4 \times \lambda_0/14 \times \lambda_0/7$	5%	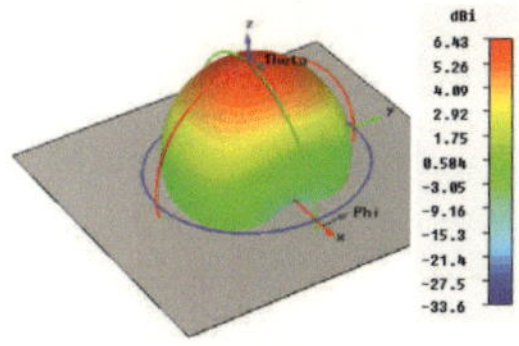	6.4	95%

6. Conclusion

To conclude, an overview of classical antennas with their miniaturization techniques has been presented and detailed in this chapter while mentioning a lot of literature references. Classical wire antennas as monopoles present good impedance bandwidth, but they remain too large to be integrated inside last generations of mobile devices. Planar antennas have the advantage to be generally low profiles and thus easier to be integrate. However, patch antennas or planar inverted F antennas have maximum gains relative to the vertical axis. Thus, wire patch antenna presents a good alternative since it radiates as a dipole antenna and is significantly smaller. Concerning the dielectric resonator antennas, they can be miniature and can resonate and be matched on different frequency by creating some partial boundary condition [50].

Author details

L. Huitema and T. Monediere

University of Limoges, Xlim Laboratory, France

References

[1] Hansen, R.C.;, "Fundamental limitations in antennas," *Proceedings of the IEEE*, vol.69, no.2, pp. 170-182, Feb. 1981

[2] McLean, J. S.;, "A re-examination of the fundamental limits on the radiation Q of electrically small antennas," *Antennas and Propagation, IEEE Transactions on*, vol.44, no.5, pp.672, May 1996

[3] Collin, R.; Rothschild, S.;, "Evaluation of antenna Q," *Antennas and Propagation, IEEE Transactions on*, vol.12, no.1, pp. 23-27, Jan 1964

[4] Huitema, L.; Sufyar, S.; Delaveaud, C.; D'Errico, R.;, "Miniature antenna effect on the ear-to-ear radio channel characteristics," *Antennas and Propagation (EUCAP), 2012 6th European Conference on*, vol., no., pp.3402-3406, 26-30 March 2012

[5] McLean, J. S.;, "A re-examination of the fundamental limits on the radiation Q of electrically small antennas," *Antennas and Propagation, IEEE Transactions on*, vol.44, pp. 672, May 1996

[6] Geyi, W.; Jarmuszewski, P.; Qi, Y.;, "The Foster reactance theorem for antennas and radiation Q," *Antennas and Propagation, IEEE Transactions on*, vol.48, no.3, pp.401-408, Mar 2000

[7] Best, S.R.;, "Bandwidth and the lower bound on Q for small wideband antennas," *Antennas and Propagation Society International Symposium 2006, IEEE*, vol., no., pp. 647-650, 9-14 July 2006

[8] Kan, H.K.; Waterhouse, R.B.;, "Size reduction technique for shorted patches," *Electronics Letters*, vol.35, no.12, pp.948-949, 10 Jun 1999

[9] Liebendorfer, et al., "Wireless LAN diversity antenna system for PCMCIA card integration," *Vehicular Technology Conference, 1997, IEEE 47th*, vol.3, no., pp.2022-2026 vol. 3, 4-7 May 1997

[10] Targonski, S.D.; et al., "Design of wide-band aperture-stacked patch microstrip antennas," *Antennas and Propagation, IEEE Transactions on*, vol.46, no.9, pp.1245-1251, Sep 1998

[11] Kossiavas, G.; Papiernik, A.; Boisset, J.P.; Sauvan, M.;, "The C-patch: a small microstrip element," *Electronics Letters*, vol.25, no.4, pp.253-254, 16 Feb. 1989

[12] Jui-Han Lu; Chia-Luan Tang; Kin-Lu Wong;, "Novel dual-frequency and broad-band designs of slot-loaded equilateral triangular microstrip antennas," *Antennas and Propagation, IEEE Transactions on*, vol.48, no.7, pp.1048-1054, Jul 2000

[13] Salonen, P.; et al., "Single-feed dual-band planar inverted-F antenna with U-shaped slot," *Antennas and Propagation, IEEE Transactions on*, vol.48, no.8, pp. 1262-1264, Aug 2000

[14] Ramirez, R.R.; Elsadek, H.; Jofre, L.; De Flaviis, F.;, "Reduced size single and dual band linear polarized microstrip antennas for mobile communications," *Antennas and Propagation Society International Symposium, 2001. IEEE*, vol.4, no., pp.450-453 vol.4, 2001

[15] Waterhouse, R.;, "Small microstrip patch antenna," *Electronics Letters*, vol.31, no.8, pp. 604-605, 13 Apr 1995

[16] Waterhouse, R.B.; et al., "Design and performance of small printed antennas," *Antennas and Propagation, IEEE Transactions on*, vol.46, no.11, pp.1629-1633, Nov 1998

[17] Colburn, J.S.; et al., "Patch antennas on externally perforated high dielectric constant substrates," *Antennas and Propagation, IEEE Transactions on*, vol.47, pp.1785-1794, 1999

[18] Niamien, C.; Collardey, S.; Sharaiha, A.; Mahdjoubi, K.;, "Compact Expressions for Efficiency and Bandwidth of Patch Antennas Over Lossy Magneto-Dielectric Materials," *Antennas and Wireless Propagation Letters, IEEE*, vol.10, no., pp.63-66, 2011

[19] Ferrero, F.; Chevalier, A.; Ribero, J.M.; Staraj, R.; Mattei, J.L.; Queffelec, Y.;, "A New Magneto-Dielectric Material Loaded, Tunable UHF Antenna for Handheld Devices," *Antennas and Wireless Propagation Letters, IEEE*, vol.10, no., pp.951-954, 2011

[20] Mattei, J.-L.; Huitema, L.; et al., "Suitability of Ni-Zn Ferrites Ceramics With Controlled Porosity as Granular Substrates for Mobile Handset Miniaturized Antennas," *Magnetics, IEEE Transactions on*, vol.47, no.10, pp.3720-3723, Oct. 2011

[21] L. Huitema, M. Hajj, T. Monédière, D. Souriou, A. Chevalier, J-L. Mattei, P. Queffelec, "*Overview of Reconfigurable and Compact Antennas Using a Magneto-dielectric Material*", Progress In Electromagnetics Research Symposium, Marrakesh, Morocco, 20-23 March 2011

[22] Huitema, L.; et al., "Compact and Multiband Dielectric Resonator Antenna With Pattern Diversity for Multistandard Mobile Handheld Devices," *Antennas and Propagation, IEEE Transactions on*, vol.59, no.11, pp.4201-4208, Nov. 2011

[23] Libo Huang; et al., "Electrically Tunable Antenna Design Procedure for Mobile Applications," *Microwave Theory and Techniques, IEEE Transactions on*, vol.56, pp. 2789-2797, Dec. 2008

[24] Li, Y., Zhang, Z., Chen, W., Feng, Z. and Iskander, M. F. (2010), A compact DVB-H antenna with varactor-tuned matching circuit. Microw. Opt. Technol. Lett., 52: 1786–1789

[25] Huitema, L.; Reveyrand, T. et al., "A compact and reconfigurable DVB-H antenna for mobile handheld devices," *Antennas and Propagation (EUCAP), Proceedings of the 5th European Conference on*, vol., no., pp.1314-1317, 11-15 April 2011

[26] Huitema, L.; Sufyar, S.; Delaveaud, C.; D'Errico, R., "Miniature antenna effect on the ear-to-ear radio channel characteristics," *Antennas and Propagation (EUCAP), 2012 6th European Conference on*, vol., no., pp.3402,3406, 26-30 March 2012

[27] Huitema, L.; Koubeissi, M.; Decroze, C.; Monediere, T.;, "Ultrawideband Dielectric Resonator Antenna for DVB-H and GSM Applications," *Antennas and Wireless Propagation Letters, IEEE*, vol.8, no., pp.1021-1027, 2009

[28] J.D. Kraus « *Antennas* » Mc Graw-Hill book company, New York, USA, 1950, pp. 127-148

[29] C.A Balanis, *Antenna Theory: Analysis and Design*, Second Edition, John Wiley & Sons, 1997

[30] J.D. Kraus, *Microstrip antenna,* Artech House, Dedham, Massachusetts, USA, 1980, pp. 2-84

[31] K. Fujimoto, A. Henderson, K. Hirasawa, J.R. james, *Small antennas*, England : Research studies press, 1987, distribued by Wiley & Sons, pp.116-127

[32] Wunsch, A.D.; Sheng-Pin Hu;, "A closed-form expression for the driving-point impedance of the small inverted L antenna," *Antennas and Propagation, IEEE Transactions on*, vol.44, no.2, pp.236-242, Feb 1996

[33] Bin Yuan; Xiu Wang; Nan Zheng; Yangbang Chen;, "An inverted-L antenna based on finite ground plane with zigzag structure," *Microwave Conference Proceedings (APMC), 2010 Asia-Pacific*, vol., no., pp.2025-2028, 7-10 Dec. 2010

[34] Oh, K.; Hirasawa, K.;, "A dual-band inverted-L-folded-antenna with a parasitic wire," *Antennas and Propagation Society International Symposium, 2004. IEEE*, vol.3, no., pp. 3131-3134 Vol.3, 20-25 June 2004

[35] Qi, D., Li, B. and Liu, H., “Compact triple-band planar inverted-F antenna for mobile handsets,” *Microwave and Optical Technology Letters*, Vol. 41, No. 6, pp. 483-486, June 2004

[36] Haapala, Paivi; Vainikainen, Pertti;, "Helical antennas for multi-mode mobile phones," *Microwave Conference, 1996. 26th European*, vol.1, no., pp.327-331, 6-13 Sept. 1996

[37] Kuo-Liang Wu, Guan-Yu Chen, Jwo-Shiun Sun, Cheng-Hung Lin, Kwong-Kau Tiong, and Y. D. Chen “The Helical Antenna for Handset Design and Phantom Effect” Progress In Electromagnetics Research Symposium, PIERS 2008 in Hangzhou, CHINA, 24-28 March, 2008

[38] Howell, J., "Microstrip antennas," *Antennas and Propagation Society International Symposium, 1972*, vol.10, no., pp.177,180, Dec 1972

[39] Colburn, J.S.; Rahmat-Samii, Y.;, "Patch antennas on externally perforated high dielectric constant substrates," *Antennas and Propagation, IEEE Transactions on*, vol.47, no. 12, pp.1785-1794, Dec 1999

[40] L. J. Chu, “Physical limitations on omni-directional antennas”, J. Appl. Phys., 1948, 19, pp. 1163–1175.

[41] Hansen, R. C.; and Burke, M.,; “Antennas with magneto-dielectrics,” *Microwave and Optical Technology Letters*, vol. 26, pp.75–78, July 2000

[42] Niamien, C.; Collardey, S.; Sharaiha, A.; Mahdjoubi, K.;, "Compact Expressions for Efficiency and Bandwidth of Patch Antennas Over Lossy Magneto-Dielectric Materials," *Antennas and Wireless Propagation Letters, IEEE*, vol.10, no., pp.63-66, 2011

[43] Ching-Wei Ling; Chia-Yu Lee; Chia-Lun Tang; Shyh-Jong Chung;, "Analysis and Application of an On-Package Planar Inverted-F Antenna," *Antennas and Propagation, IEEE Transactions on*, vol.55, no.6, pp.1774-1780, June 2007

[44] C. Delaveaud, P. Leveque, B. Jecko, "New kind of microstrip antenna: the monopolar wire-patch antenna," *Electronics Letters*, vol.30, no.1, pp.1-2,6 Jan. 1994

[45] S. Sufyar, C. Delaveaud " A miniaturization technique of a compact omnidirectional antenna ", *Radioengineering*, December 2009, Volume 18, Number 4

[46] S. Sufyar, C. Delaveaud, R. Staraj, "A Frequency Agility Technique on a Miniature Omnidirectional Antenna", *4th European Conference on Antennas and Propagation (EUCAP 10)*, 12-16 April, Barcelona

[47] HUITEMA, L.; DELAVEAUD, C.; D'Errico, R., "Impedance and Radiation Measurement Methodology for Ultra Miniature Antennas," *Antennas and Propagation, IEEE Transactions on*, vol.62, no.7, July 2014

[48] A. Petosa, "Dielectric Resonator Antenna Handbook", Artech House, Boston/London, 2007

[49] Petosa, A.; Simons, N.; Siushansian, R.; Ittipiboon, A.; Cuhaci, M.;, "Design and analysis of multisegment dielectric resonator antennas," Antennas and Propagation, IEEE Transactions on, vol.48, no.5, pp.738-742, May 2000

[50] L. Huitema, M. Koubeissi, C. Decroze, T. Monediere, "Ultrawideband Dielectric Resonator Antenna for DVB-H and GSM Applications" IEEE Antennas and Wireless Propagation letter, vol. 8, pp. 1021-1027, 2009

Antennas for Body Centric Wireless Communications at Millimeter Wave Frequencies

Nacer Chahat, Maxim Zhadobov and
Ronan Sauleau

Additional information is available at the end of the chapter

http://dx.doi.org/10.5772/58816

1. Introduction

Over the past few years, wearable antennas for body centric wireless communication systems have been increasingly gaining attention. Body area networks (BANs) are wireless communication systems that enable communications between wearable and/or implanted electronic devices.

Such systems are of great interest for various applications such as healthcare, entertainment, military, identification systems, sport, smart home, and space [1, 2]. Indeed, portable devices, incorporating antennas close to the human body, have been used for many decades by the military. Nowadays, in order to increase overall effectiveness of soldiers on the battlefield, works are undertaken to integrate wireless systems to all equipment, such as weapons, sighting systems, helmet, and so on. Besides, wearable or implanted sensors increase the ability of doctors to monitor their patients at long distance and in real time. This monitoring capability is also used in sport applications and in rescue worker interventions.

Recently, studies carried out at millimeter waves have grown rapidly. This is due to the fact that many advantages can be found in operating such systems at millimeter waves compared to microwaves. First, because of the large available spectrum (7 GHz worldwide), very high data rates can be reached (up to 5 Gb/s) [3]. Second, it provides a high level of security and low interference with adjacent networks [4]. Finally, compared to on-body devices operating at microwaves, the size of similar millimeter-wave systems is significantly reduced.

Hence, the use of millimeter wave systems for BANs will have a high impact, in particular in the defense sector (Fig. 1), where communications emanating from a dismounted soldier leads to detection, location and vulnerability to enemy attack. The high atmospheric attenuation in the 60-GHz band will lead to much higher levels of security against detection, interception and jamming. Fig. 1 illustrates a scenario of soldier-to-soldier communications for covert battlefield operation where co-located soldiers are wirelessly networked to allow high-speed communications within a cluttered urban warfare environment. Besides, every soldier is equipped with advanced technology significantly improving situational awareness, lethality and survivability such as GPS, helmet mounted display, RADAR bullet detector, etc.

In addition, millimeter wave BANs will also benefit civilian sectors such as healthcare, personal entertainment, sports training, and emergency services. In hospital, clinics, entertainment venues, and public transport, there is a need to relay personalized data to and from individuals, in confined areas, or in crowds, and the high frequency and highly directive beams from small millimeter-wave antennas will reduce interference between users and other communication equipment.

The aim of this book chapter is to provide a review of recent progresses and outstanding challenges in the field of antennas for body-centric communication at millimeter waves.

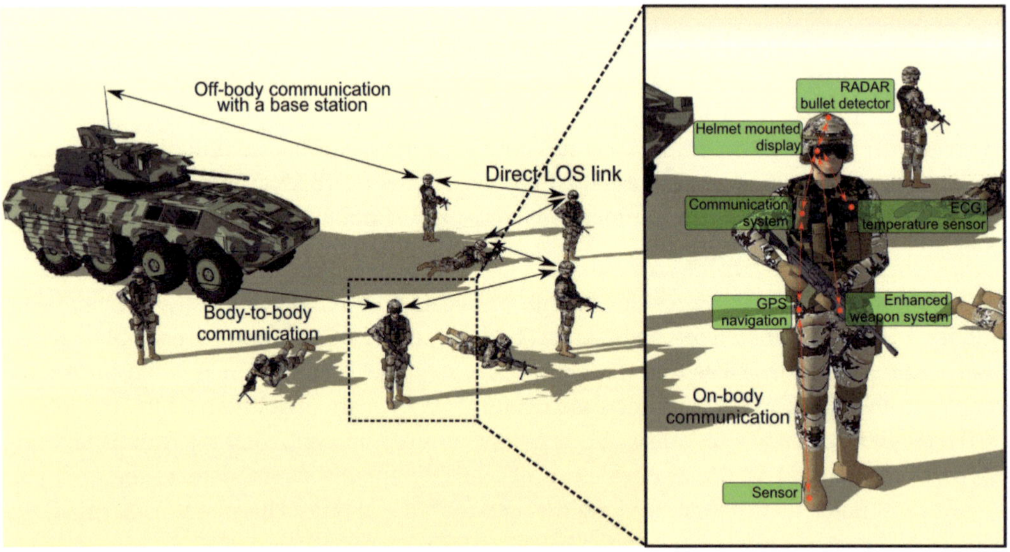

Figure 1. Soldier-to-soldier communications for covert battlefield operations. The black arrows represent some possible wireless links allowing data transfer from one soldier to another.

2. Electromagnetic properties and modeling of the human body

In this Section, the electromagnetic properties and modeling of the human body are investigated. First, to study the interaction of millimeter waves and the human body, the skin dielectric properties are carefully characterized. Then, the influence of the antenna feeding is investigated. Then, as the dielectric properties of the skin have been assessed, a numerical model of the human body is introduced using a Debye model. Finally, a semi-solid phantom is introduced for antenna measurement in close proximity to the body.

2.1. Interaction of millimeter waves with the human body

The primary biological targets of 60-GHz radiations are the skin and eyes. Exposure of the eyes leads to the absorption of the EM energy by the cornea characterized by a free water content of 75% and a thickness of 0.5mm. Ocular lesions have been found after high-intensity exposure of the eye (3W/cm², 6min) [5]. However, studies performed at 60 GHz (10mW/cm², 8h) demonstrated no detectable physiological modifications [6], indicating that millimeter waves act on the cornea in a dose-dependent manner.

Hereafter we will essentially consider the interactions with the skin as it covers 95% of the human body surface. From the EM viewpoint, human skin can be considered as an anisotropic multilayer dispersive structure made of three different layers, namely, epidermis, dermis, and subcutaneous fat layer (Fig. 2). The skin also contains capillaries and nerve endings. It is mainly composed of 65.3% of free water, 24.6% of proteins, and 9.4% of lipids [7].

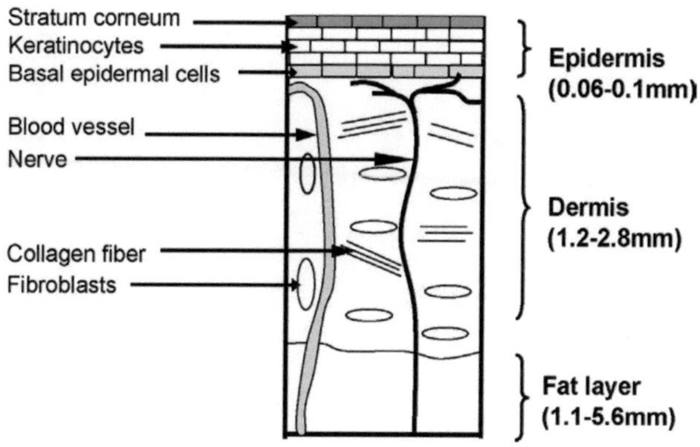

Figure 2. Schematic representation of the skin structure.

Knowledge of the dielectric properties of the skin is essential for the determination of the reflection from, transmission through, and absorption in the body, as well as for EM modeling. In contrast to frequencies below 20 GHz, existing data on the permittivity of tissues in the millimeter-wave band are very limited [8]-[11] due to some technical difficulties. In the 10–100 GHz range, the dispersive dielectric properties of the skin and biological solutions are primarily related to the rotational dispersion of free water molecules. In particular, high losses are related to the free water relaxation with the peak at 23 GHz at 33°C.

In contrast to frequencies below 20 GHz, the already-existing data on the relative permittivity of human tissues at millimeter waves are very limited. In addition, the results reported so far in the literature strongly depend on the measurement technique, the sample type (*in vivo* or *in vitro* study) and other experimental conditions such as skin temperature, location on the body and thickness of different skin layers.

Table 1 provides a summary of the data previously reported at 60 GHz. These results show that the literature data vary significantly from one study to another depending on the sample type. Besides, since the skin consists of approximately 65% of free water [7], its complex permittivity is strongly dispersive and temperature-dependent; this should be also taken into account for definition of an accurate skin permittivity model.

To validate our measurement technique and obtain reference data for the skin-equivalent phantom, we performed a measurement campaign on a group of volunteers using two different techniques: open-ended coaxial slim probe [12] and a new method based on heating kinetics [13]. A very good agreement is demonstrated between our measurements and Gabriel [10] and Alekseev [11] data as shown in Fig. 3.

Reference	Complex permittivity ε^*	T, °C	Method	Sample type
Gandhi *et al.* [8]	$8.89 - j13.15$	37±0.5	E	*In vitro*
Alabaster *et al.* [9]	$9.9 - j9.0$	23	M	*In vitro*
Gabriel *et al.* [10] "wet skin"	$10.22 - j11.84$	37	E	*In vitro*
Gabriel *et al.*[10] "dry skin"	$7.98 - j10.90$	32.5±0.5	E	*In vivo*
Alekseev *et al.* [11]	$8.12 - j11.14$	32.5±0.3	M	*In vivo*
Chahat et al. [12]	$8.02 - j10.5$	32.5±0.5	M	*In vivo*
Chahat *et al.* [13]	$8.4 - j10.96$	32.5±0.5	M	*In vivo*

E=Extrapolation. M=Measurement. T=theoretical value.

Table 1. Overview of the skin dielectric properties at 60 GHz.

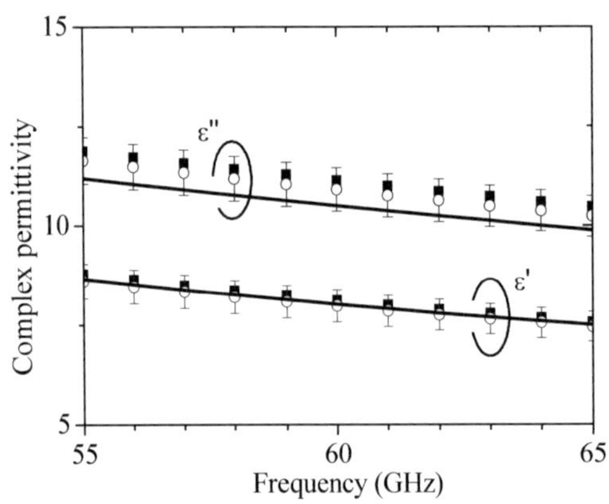

Figure 3. Comparison of our experimental result for the wrist skin permittivity (——) with Gabriel *et al.* (dry skin) (■) and Alekseev *et al.* (○) models. Error bars represent ±5% deviations around Gabriel's reference values.

http://dx.doi.org/10.5772/58816

2.2. Numerical skin-equivalent phantom

Taking into account the very shallow penetration of millimeter waves into the skin (typically 0.5 mm at 60 GHz), using homogeneous skin-equivalent phantoms provides accurate results for the antenna / human body interaction evaluation as well as for the propagation channel characterization [28]. For the broadband analysis, dispersive models can be used. Debye model with a single relaxation time τ equal to that of free water at the same temperature was demonstrated to provide a good accuracy for modeling the experimental permittivity data in the considered frequency range [12]:

$$\varepsilon^* = \varepsilon_\infty + \frac{\Delta\varepsilon}{1 + j\omega\tau} + \frac{\sigma}{j\omega\varepsilon_0}. \tag{1}$$

In this equation, $\omega=2\pi f$, f [Hz] is the frequency, $\Delta\varepsilon=\varepsilon_s-\varepsilon_\infty$ is the magnitude of the dispersion of the free water fraction of skin, ε_s is the permittivity at $\omega\tau<<1$, ε_∞ is the optical permittivity, $\varepsilon_0=8.85\ 10^{-12}$ F/m, and σ [S/m] is the ionic conductivity. The optimized parameters that fit to the measured permittivity in the 55-65 GHz range are the following: $\varepsilon_\infty=4.1$, $\varepsilon_s=34.8$, $\tau=6.9 \times 10^{-12}$ s, and $\sigma=0.7$ S/m [12]. This model allows an accurate representation of typical broadband dielectric properties of dry skin in the numerical modeling.

2.3. Experimental skin-equivalent phantom

2.3.1. Composition

The main components employed for the fabrication of a homogeneous semi-solid skin-equivalent phantom are the following:

- Deionized water. Water is the main constituent of the phantom because it is also the main skin component. It primarily determines the dispersive behavior of the phantom.
- Agar. It is employed for the retention of self-shaping, and its contribution to the phantom dielectric properties is negligible for small concentrations (typically below 4%).
- Polyethylene powder. It is used to tune the real and imaginary parts of the phantom permittivity.
- TX-151. Since the agar and polyethylene powder cannot be mixed directly, the viscosity is increased using TX-151.
- Sodium azide (NaN3). It serves as a preservative.

2.3.2. Fabrication procedure

The fabrication steps are the following. Deionized water, sodium azide, and agar are mixed in a kettle and heated on a stove, while the mixture is continuously stirred. When this liquid starts boiling, heating is stopped. TX-151 is sprinkled into the liquid and quickly mixed. Then the polyethylene powder is added into the stirred liquid. Finally, the obtained mixture is poured into a mold and cooled in the same container for a few hours to room temperature for solidification. Using alginate gel powder, molds with realistic body-specific shapes can be manufactured for the phantom fabrication as illustrated in Fig. 4.

Particular attention should be paid to the following critical points. First, to avoid variations of dielectric properties from one phantom to another, the room temperature should remain identical (in our case 20±1°C) during the fabrication and further measurements. Second, the type of polyethylene powder is important; we recommend using particles with an average diameter of 20μm and low density ~ 900–1100 kg/m^3. Finally, to preserve the dielectric properties of the phantom over time, it is important to avoid water evaporation since this would result in a decrease of the permittivity. This can be for instance achieved by wrapping the phantom in a plastic film. More details regarding the phantom preparation procedure can be found in [12].

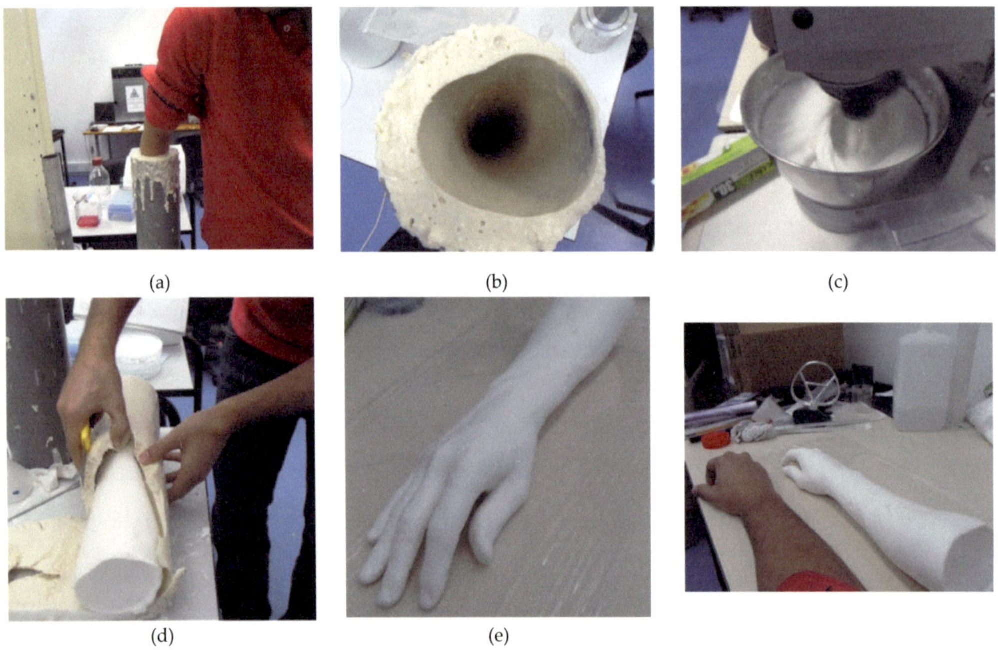

Figure 4. Skin-equivalent phantom representing an arm and a hand: (a) fabrication of an alginate mold; (b) alginate mold of a human arm; (c) fabrication of the phantom liquid; (d) the phantom is extracted after being cooled inside the mold; (e) final result of a realistic human arm phantom.

2.3.3. Dielectric properties

The measured dielectric properties of the skin-equivalent phantom and skin are compared in Fig. 5. The dielectric properties of the proposed phantom are within ±10% of the measured skin permittivity. Table 2 compares the dielectric properties of the phantom measured using the coaxial probe and the heating kinetics technique [13] to those of the reference values

http://dx.doi.org/10.5772/58816

provided by Gabriel *et al.* [10]. The measured data are in excellent agreement with the reference data and demonstrate that this phantom can be used for antenna measurement, on-body propagation, and dosimetric studies.

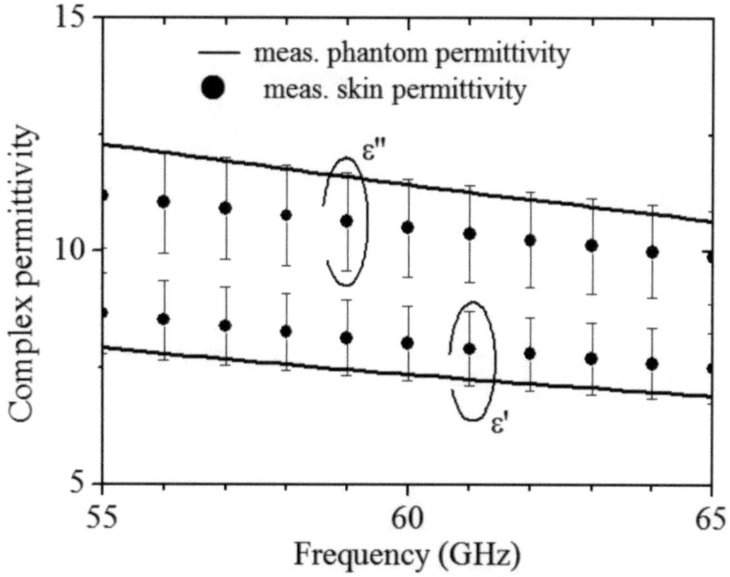

Figure 5. Dielectric properties of the skin-equivalent phantom compared to those of skin [12]. Error bars represent ±10% of the measured skin permittivity.

	ε*	***Δε****	***R***	***δ (mm)***
Reference value (Gabriel *et al.*) [10]	7.98 – *j*10.9	–	0.38	0.48
Phantom (coaxial probe)	7.4 – *j*11.4	7.3% - *j* 4.6%	0.39	0.45
Phantom (heating kinetics)	8.3 – j10.8	4% - *j* 0.9%	0.38	0.49

Table 2. The dielectric properties of the proposed phantom (using two different techniques) compared to those of the reference data provided by Gabriel *et al.* [10]. *Δε** is the error relative to Gabriel *et al.* data.

2.3.4. Validation

To further confirm the reliability of this phantom, we performed SAR measurement using a high-performance thermal imaging camera (FLIR SC500, FLIR Systems, Wilsonville, OR, USA) and the measurement set-up shown in Fig. 6a. The SAR assessment methodology is described in Fig. 6b. The temperature dynamic, recorded using the IR camera, is fitted to the one-dimensional bio-heat transfer equation [12]. The fitting procedure is performed by minimizing the standard deviation value varying the incident power density (IPD). Once the IPD value has been determined, the SAR can be found (Fig. 6b). The simulated and measured SAR results are in excellent agreement (Fig. 6) which confirms the accuracy of this phantom.

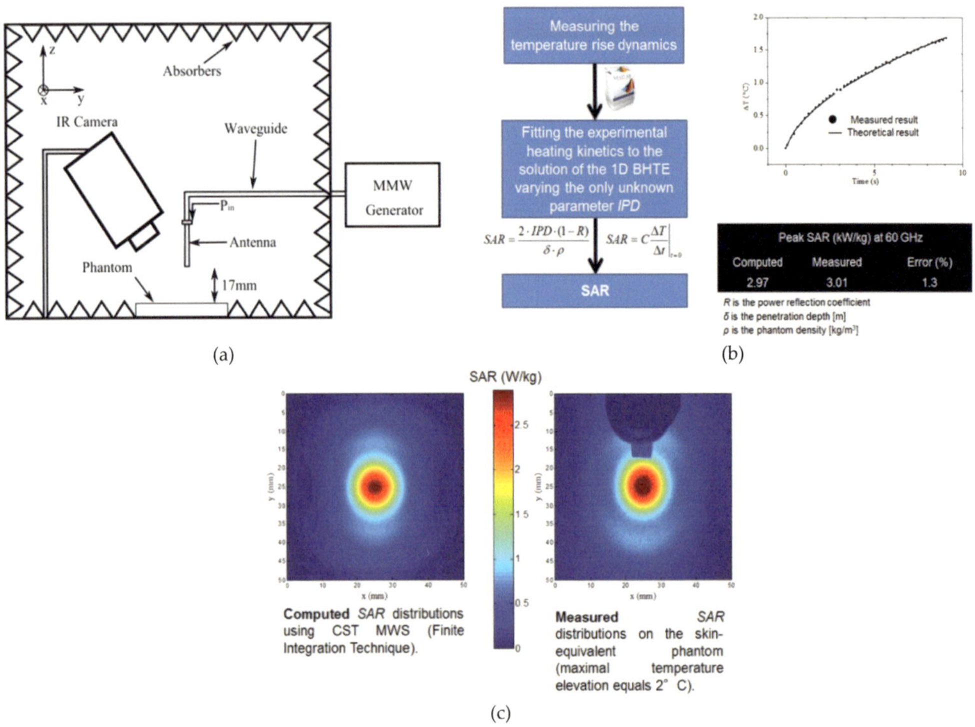

Figure 6. (a) SAR measurement set-up. (b) SAR assessment methodology from the temperature rise. (c) Simulated and measured SAR results.

3. Antennas for off-body communications at millimeter-waves

At microwaves, it is widely accepted that antennas placed in close proximity to a lossy medium experience strong power absorption, radiation pattern distortion, shift in resonance frequency, and changes in the input impedance, e.g. [1],[19]-[21]. Therefore, when placed close to the human body, wearable antennas need to be designed to operate in a robust way so that the influence of the body on the antenna performance is minimized. Patch antennas have been identified as one of the best solutions for off-body communications [1]. These are simple and low-cost structures, and their radiation at broadside allows maximizing radiations at the opposite side of the human body while reducing radiation towards the body.

At millimeter waves, the electromagnetic coupling between antennas and the human body as well as possible perturbations of antenna characteristics due to the body remain almost unexplored. In addition, in this frequency range, a particular attention must be paid to the power absorbed in the body since this absorption is very localized.

In this Section, the interactions between the human body and millimeter wave antennas, optimized for off-body communications, are studied numerically and experimentally. First, requirements for wearable antennas for off-body communications are briefly outlined. Then, the influence of the antenna feeding is investigated. Then, a four-patch antenna array is designed and characterized numerically and experimentally both in free space and on the skin-equivalent phantom described in the previous section. SAR and incident power density distributions on the phantom are determined using the methodology presented in [12]. Finally, in order to study the capabilities of the integration into textiles, a similar four-patch antenna array is designed and fabricated on a fabric.

3.1. Antenna requirements for off-body communications

Wearable antennas have to be as compact as possible to be integrated with the transceiver. They have to be efficient with minimal power absorption inside the human body that behaves as a highly lossy dispersive dielectric material at millimeter waves. The antennas also have to be light weight and, in some particular cases, conformable to the human body shape. Because of the high atmospheric attenuation at 60 GHz and limitations on the radiated power, medium-gain antennas (~12dBi) are often required [14]. Indeed, in controlled environments, line-of-sight (LOS) channels can be efficiently exploited using medium-gain passive antennas, whereas directive beam steering antennas are desirable for non-line-of-sight (NLOS) channels so as to comply with the power link budgets [14]-[18]. In our studies, we only consider LOS scenarios and thus restrict our consideration to passive medium-gain antennas.

3.2. Influence of the antenna feeding

The influence of the antenna feeding is investigated when the antenna is placed on the human body. At lower frequencies, patch antennas have been presented as the best solution for off-body communications. However, at millimeter waves, the influence of spurious waves due to the feeding lines on radiating patterns cannot be neglected. That is why, multilayer antenna designs are generally considered in order to overcome this issue.

The interaction with the human body and two types of patch antennas is studied: (1) a linearly-polarized antenna and (2) a linearly-polarized aperture coupled patch antenna. These antennas are printed on a 0.127mm-thick RT Duroid 5880 substrate (h=127 μm, ε_r=2.2, $tan\delta$=0.003).

3.2.1. Microstrip patch antenna

A simple patch antenna is optimized to achieve a maximum gain at 60 GHz. The dimensions are given in Fig. 7. The reflection coefficient and radiation patterns are studied numerically in free space and on the human body when the antenna is placed at 1mm above the phantom. For the numerical modeling, a parallelepipedic 10 × 100 × 100 mm^3 phantom is used and a Debye model has been used to express the complex permittivity ε^* of the skin-equivalent phantom (see Section 2).

The reflection coefficient is very slightly affected by the human body (Fig. 8) and the radiation pattern remains stable at the opposite side of the human body, whereas the backward radiations are highly reduced in H-plane (Fig. 9). These results demonstrate that microstrip patch antennas are only slightly sensitive to the human body proximity at 60 GHz.

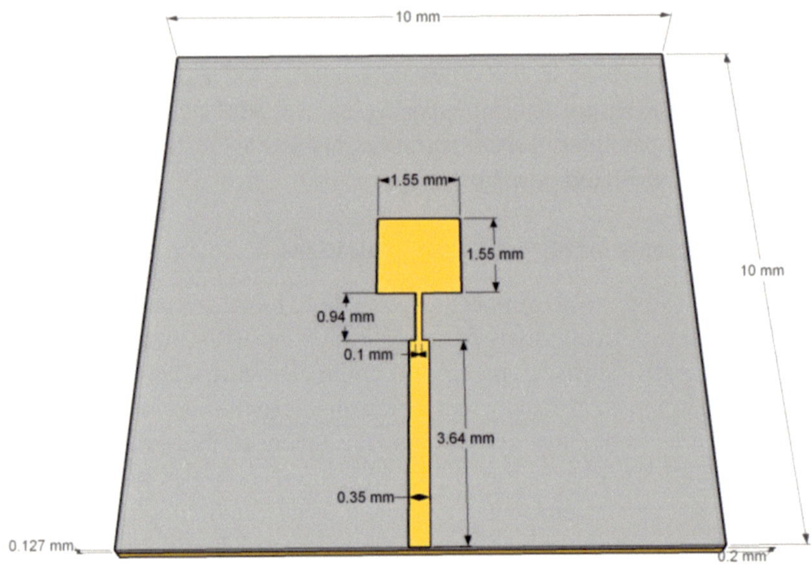

Figure 7. Microstrip patch antenna at 60 GHz.

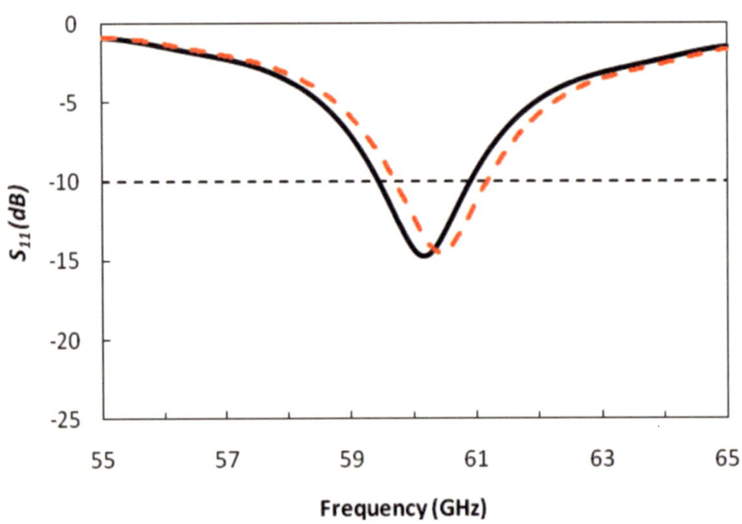

Figure 8. Simulated reflection coefficient of the microstrip patch antenna. —— In free space. — — On the skin-equivalent phantom.

http://dx.doi.org/10.5772/58816

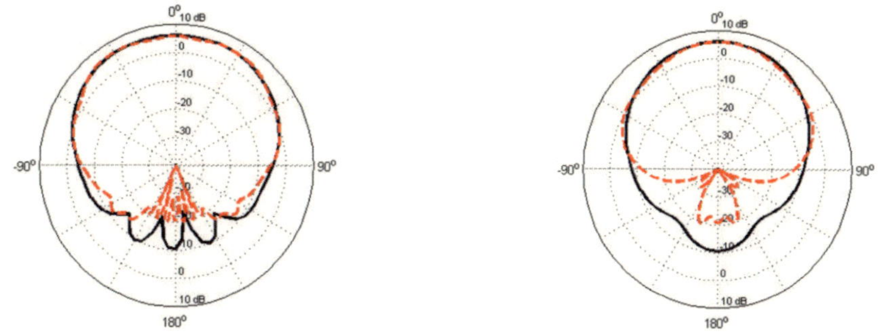

(a) *E*-plane (b) *H*-plane

Figure 9. Simulated radiation pattern of the microstrip patch antenna. —— In free space. — — On the skin-equivalent phantom.

3.2.2. Aperture coupled patch antenna

Fig. 10 shows the configuration of the aperture-coupled patch antenna (ACPA). The slot is optimized to 0.26 × 1 mm² for maximum coupling with a stub length of 0.34 mm. In order to consider the easiness of implementation, a 0.2-mm-thick ground plane is employed. The antenna consists of a patch with optimized dimension of 1.33×1.24 mm² on a 0.127-mm-thick RT Duroid 5880 substrate. Low thickness and low-permittivity substrate are used for reducing surface waves.

The reflection coefficient S_{11} (Fig. 11) and radiation patterns (Fig. 12) are investigated in free space and on the skin-equivalent phantom (antenna/body spacing h=1mm). It is worthwhile to note that the S_{11} is even less affected by the human body proximity for the ACPA. However, as far as the radiation pattern is concerned, the backward radiations are highly reduced (i.e. by at least 10 dB). This demonstrates that absorptions inside the body are higher for the ACPA and the SAR should be carefully studied. The gain is very slightly increased on the human body (Table 3).

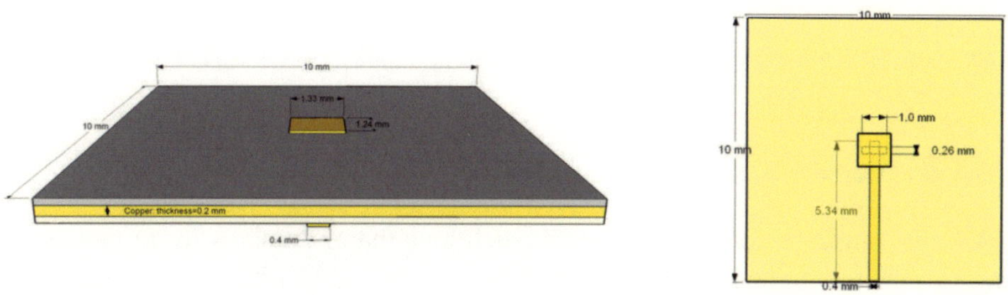

(a) (b)

Figure 10. Aperture coupled patch antenna. (a) 3D and (b) 2D schematic representation of the antenna model and dimensions.

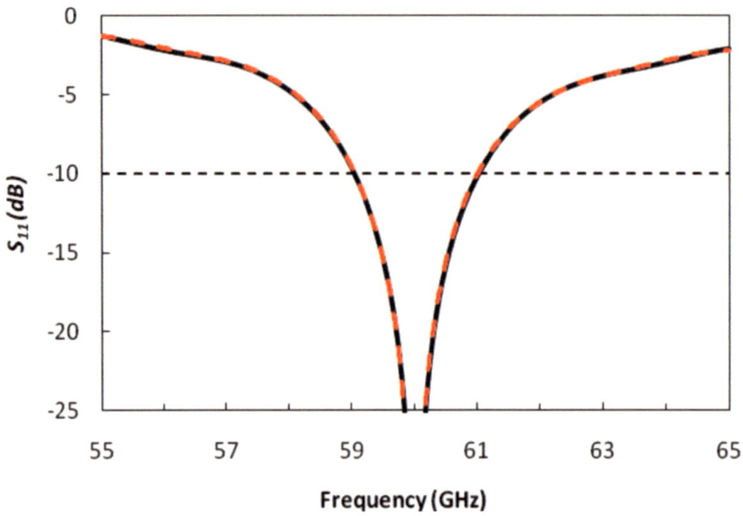

Figure 11. Simulated reflection coefficient of the aperture coupled patch antenna. —— In free space. — — On the skin-equivalent phantom.

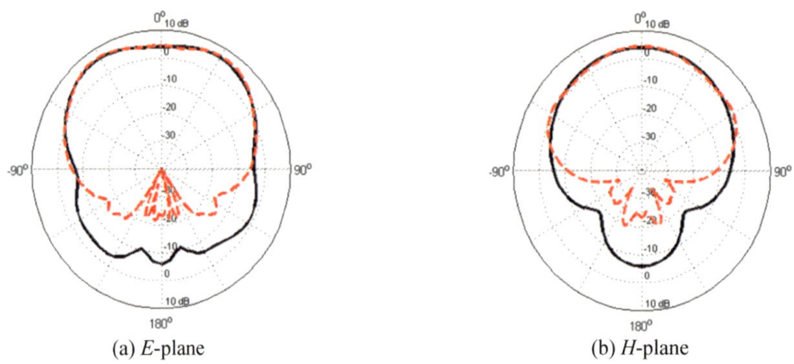

Figure 12. Simulated radiation pattern of the aperture coupled patch antenna. —— In free space. — — On the skin-equivalent phantom.

3.2.3. Specific Absorption Rate (SAR) comparison

The SAR are compared for the microstrip patch antenna and ACPA for an antenna/body spacing *h*=1mm and for an incident power of 1W. The peak SAR obtained for the ACPA is 41 times higher compared to that obtained with the microstrip patch antenna. Therefore, it is not recommended to use ACPA because the input power would be highly limited compared to that of the microstrip patch antenna to comply with the exposure limits [22], resulting in a lower link budget (gain and efficiency remain almost equivalent for both antennas).

	Microstrip patch antenna		ACPA	
	Free space	On the phantom	Free space	On the phantom
Peak SAR (W/kg)[1]	-	279	-	11485
Peak gain (dBi)	6.01	6.03	6.22	6.70
Efficiency (%)	79.31	74.44	84.58	77.67

Table 3. Peak SAR, gain, and efficiency for the microstrip patch antenna and ACPA. [1]For an incident power of 1W

3.2.4. Conclusion

Two patch antennas have been compared numerically in free space and on a skin-equivalent phantom. For the microstrip antenna and ACPA, the influence of the human body is very weak, and their performances remain stable. However, the SAR resulting from the ACPA is 41 times higher compared to that obtained with the microstrip antenna. Therefore, it is highly recommended to avoid aperture coupled feeds. If it is necessary, the feeding line could be sandwiched between two substrates with top and bottom grounds [23].

3.3. Patch antenna array

3.3.1. Antenna model

To satisfy the criteria summarized in Section 3.1 and following the conclusions drawn in Section 3.2, a microstrip-fed four-patch single-layer antenna array has been chosen [24]. It is printed on a thin RT Duroid 5880 substrate (h=127 μm, ε_r=2.2, $tan\delta$=0.003). The layout is represented in Fig. 13a. A single rectangular patch antenna typically provides a 7 dBi gain; a 2×2 antenna array is chosen here to reach a gain of 12 dBi with about the same beamwidth in E-and H-planes. The inter-element spacing is selected to achieve a good trade-off between high gain and low side lobes. Similar 2×2 antenna arrays have already been reported in a multilayer configuration [25] or fed by a coaxial probe [26],[27], which would make them difficult to fabricate on flexible or textile substrates. Hence, here all patches are fed using a single-layer corporate feed network. The antenna is linearly-polarized along y-direction, and, for measurement purposes, it is mounted on a 3 mm-thick ground plane (Fig. 13b) to avoid significant substrate bending and to achieve an accurate and stable placement of a V-connector. In the future BAN applications, this kind of antennas is expected to be directly integrated into the clothing or wearable devices.

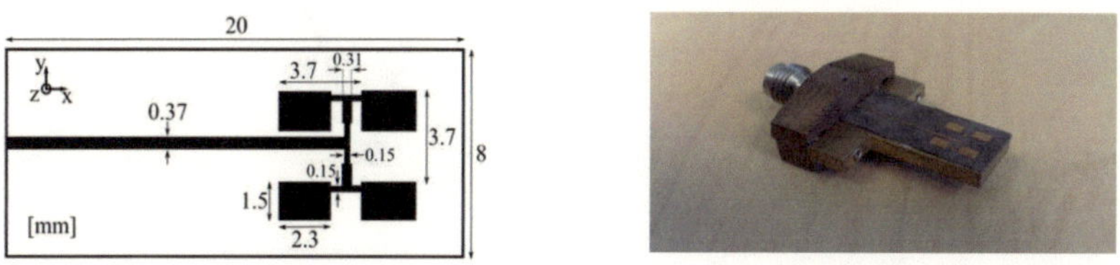

(a) (b)

Figure 13. Patch single-layer antenna array at 60 GHz [24]. (a) Schematic representation of the antenna model and dimensions. (b) Manufactured antenna with a V-connector.

3.3.2. *Antenna performance*

The antenna reflection coefficient S_{11} was measured in free space and on the skin-equivalent phantom (Fig. 14). It remains below-10 dB from 59 GHz to 65 GHz. It is clear that the skin-equivalent phantom does not affect the antenna reflection coefficient. This is not the case at microwaves where a resonance shift is usually observed.

In addition, the radiation patterns in E-and H-planes are plotted in Fig. 15 at 60 GHz. The gain was measured by the comparison method with a 20-dBi standard horn, and the directivity is found from a 3D radiation pattern measurement. It can be seen in Fig. 15 that front radiations remain very slightly affected.

The backward radiation was measured separately in both configurations. Whereas the measured level on the phantom is mainly reduced in the *E*-plane (Fig. 15c) due to the absorption and reflection, it remains very slightly affected in the *H*-plane (Fig. 15d). This could be expected since the absorption is higher when the *E*-field is parallel to the phantom surface [28]. These observations are in agreement with the calculated and measured SAR and incident power density (IPD) as shown in Fig. 16. More details regarding the measurement methodology can be found in [12].

At this frequency, the measured gains in free space and on the phantom equal 11.8 dBi (±0.3dB) and 11.9 dBi (±0.3dB), respectively. This demonstrates the small effect of the phantom presence. At 60 GHz, the measured directivity was assessed to be equal to 13.9 dBi (±0.3dB) and 14.1 dBi (±0.3dB), respectively. Comparison of the measured directivities with the measured gains leads to antenna efficiencies of 62% and 60%, respectively. This efficiency value is typical in V-band for this kind of antennas and could be further improved, for instance using a fused quartz substrate [25] instead of RT Duroid 5880. Whereas the antenna efficiency at microwaves can be strongly affected by the body presence even for patch antennas [1], it is found here that it remains stable in V-band.

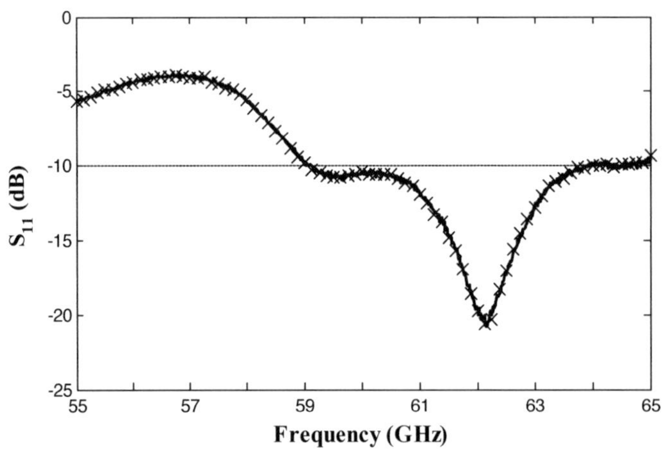

Figure 14. Measured reflection coefficient of the antenna. ——— In free space. xxx On the skin-equivalent phantom.

http://dx.doi.org/10.5772/58816

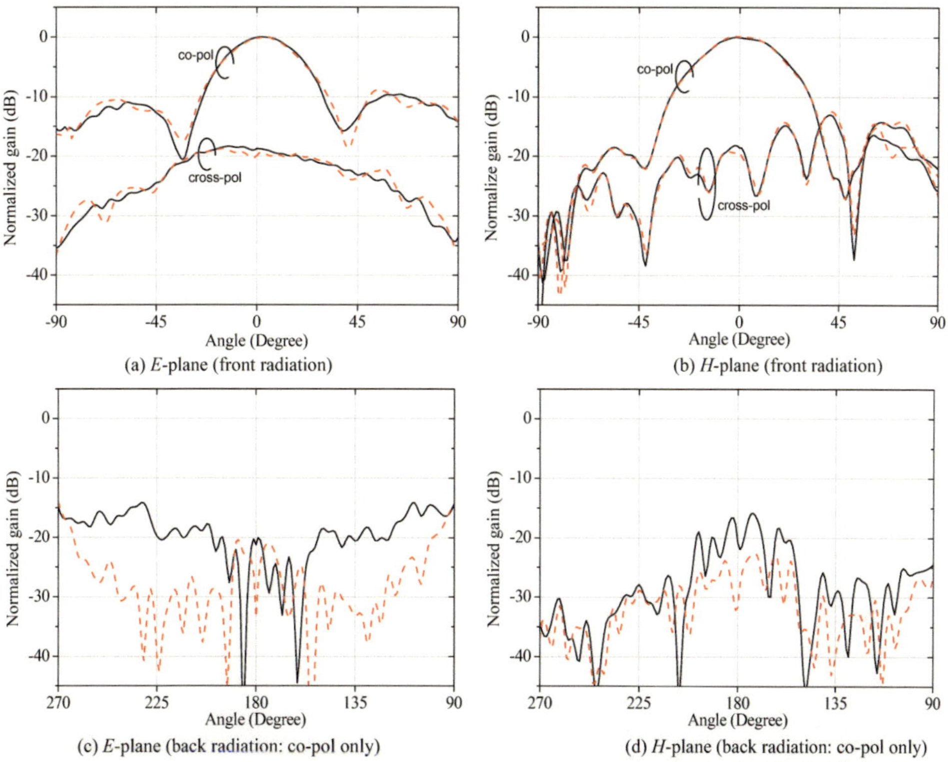

Figure 15. Measured normalized radiation patterns at 60 GHz in *E*-and *H*-planes. —— Measurement in free space. --- Measurement on the phantom.

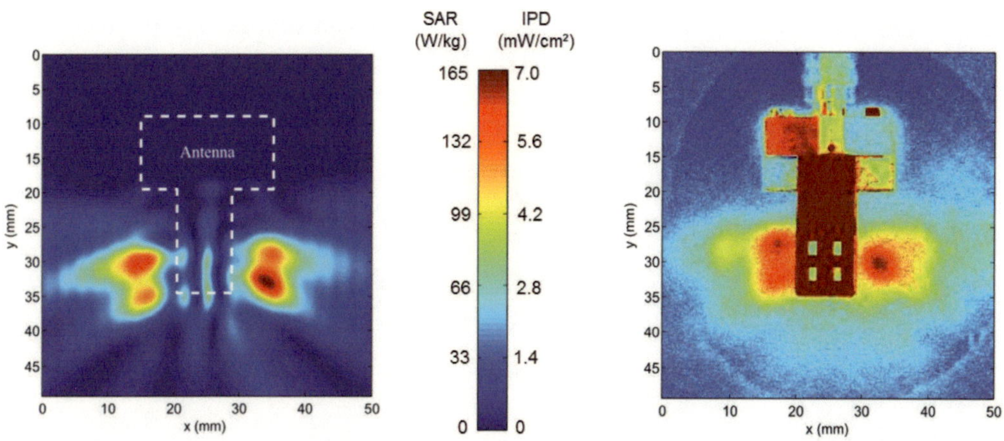

Figure 16. SAR and IPD distributions at 60 GHz. (Left) Numerical results for the antenna on the skin. (Right) Measurements on the skin-equivalent phantom. P_{in}=322 mW.

3.4. Textile antennas

Textile antennas at millimeter waves could be of great interest for many applications. However, on-textile fabrication process is very challenging at these frequencies, especially due to the roughness of the textile surface and the size of textile fibers and electrotextiles with respect to the geometrical dimensions of the metallic patterns.

It was demonstrated in [29] that commercial textiles can be used as antenna substrates at millimeter waves. Some results are presented here showing a 60-GHz textile-based antenna for off-body wireless communications with the ability to be bent and deformed into an arbitrary shape. A simple, but representative patch antenna array is fabricated using an ad-hoc manufacturing process. Compared to the antenna presented in Section 3.3, this results in a highly flexible antenna.

3.4.1. Technological fabrication process

The fabrication process of millimeter-wave textile antennas has been presented in [29] and [31] (Fig. 17). The first step (Fig. 17a) consists in placing an electrotextile layer (e.g. *ShieldIt Super*) on the lower side of the textile (ground plane), and flexible copper foil on the top side. The second step (Fig. 17b) consists in micromachining the copper foil using a laser machine with optimized laser parameters to avoid any damage of the textile substrate such as ragged or burnt edges.

Hence, using a laser machine (ProtoLaser S, LPKF, OR, USA) operating at 1064nm with a pulse duration of 7.5ns and a spot size equals 25 μm, the laser parameters were optimized. A laser fluence of 24.4 mJ/cm^2 with three cycles on the surface of the substrate has been used for the copper foil ablation (repetition rate=75 kHz, power=16.0 W) without affecting the textile substrate. These fabrication conditions lead to a geometrical accuracy of about 10 μm. It is worthwhile to underline that the accuracy reported so far with two conductive fabrics, namely knitted P130 and woven Nora fabric, is only about ±0.5mm and ±0.15mm, respectively [32]. Finally, the last step (Fig. 17c) consists in manually removing the unwanted parts of the copper foil from the surface of the textile.

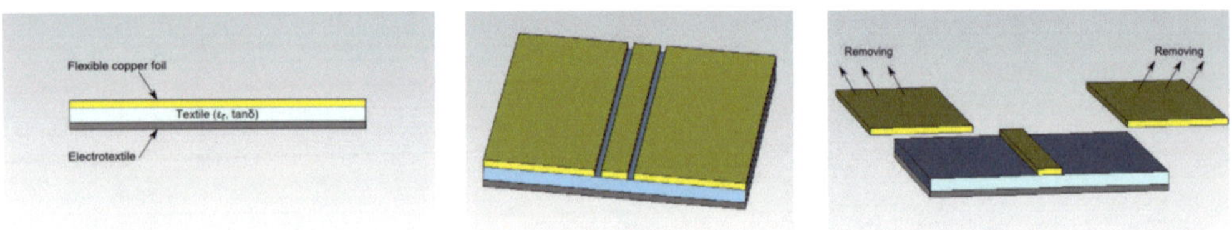

(a) Substrate preparation (b) Laser ablation (c) Removing of undesired parts

Figure 17. Main technological steps for the manufacturing of printed circuits and antennas on textiles in V-band.

Whereas in most fabrication processes reported so far, the metallic part is cut separately and then adhered to the dielectric layer, cutting out the desired pattern directly on the dielectric layer avoids additional discrepancies. Example of manufactured microstrip antennas and lines

http://dx.doi.org/10.5772/58816

are shown in Fig. 18. The devices are very flexible and the pattern quality (dimensions, sharpness of the edges) is very satisfactory (Fig. 18).

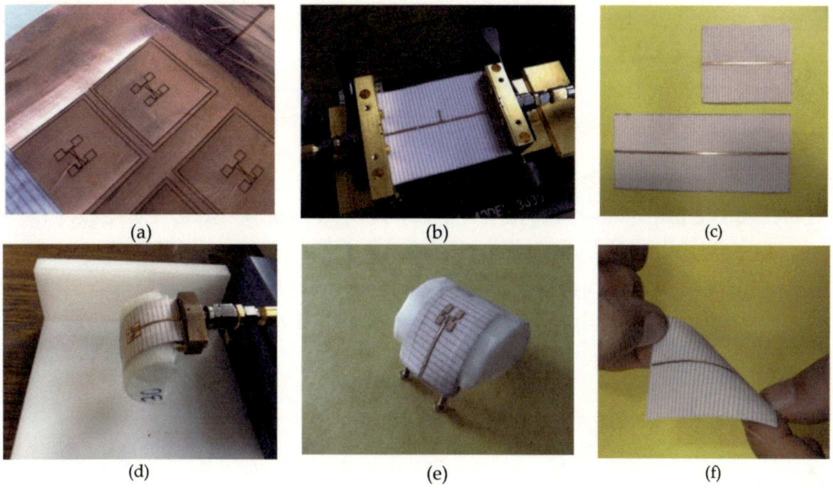

Figure 18. Examples of fabricated textile antennas and microstrip lines.

3.4.2. Textile characterization

The choice of a substrate thickness, dielectric constant ε_r and loss tangent tanδ is essential when it comes to millimeter waves in order to avoid losses and also to enhance the efficiency. The methodology employed to retrieve the dielectric properties of any textile layer is explained here. As an example, this methodology is applied to a 0.2 mm-thick cotton woven fabric extracted from a shirt. Its permittivity and loss tangent are determined in V-band as explained below. The devices under test have been manufactured using the fabrication process described in Section 3.4.1.

The characterization technique is simple and straightforward and consists in two parts:

- First, the relative permittivity is retrieved using the open-stub technique. To this end, we have designed a transmission line loaded by an open-ended parallel stub (Fig. 18b) whose length l_s is chosen to provide a resonance close to 60 GHz. The resonant frequency is measured in transmission with a V-band Anritsu universal test fixture 3680 V (Fig. 18b) connected to an Agilent 8510XF vector network analyzer (VNA). The measurement set-up has been calibrated using a full 2-port calibration procedure. The measured transmission coefficient S_{21} is represented in Fig. 19a (solid line) from 10 to 65 GHz. In simulations, the relative permittivity of the textile is tuned numerically until the theoretical S_{21} curve coincides with the measured one.

- Second, the loss tangent is estimated through a differential measurement in transmission of two matched 50-Ω microstrip lines of different lengths (Fig. 18c). This enables determination of the total insertion loss (Fig. 19b), and tanδ is found by fitting the measured and simulated data. Our experimental data show that the insertion loss of transmission lines fabricated on cotton woven fabric reaches about 1.6 dB/cm at 60 GHz, which is larger than values obtained with conventional substrates [29].

The best agreement between simulations and experiments is obtained with ε_r=2.0 and $tan\delta$=0.02. These values will be used for the antenna design. Since commercial textiles are lossy, a slight deviation in the determination of their loss-tangent would have a minor impact. Therefore, deviations due to the use of electromagnetic software are considered as acceptable.

The insertion loss of a 50-Ω microstrip line printed on a 0.2mm-thick textile is about 1.6 dB/cm, which is quite important compared to conventional substrates such as RT Duroid 5880, fused quartz and alumina [29]. However, these substrates are not as flexible as textiles. For a fair comparison, we should consider a flexible substrate such as PDMS where the insertion losses are much more important (~3 dB/cm for a 0.2mm-thick PDMS) [30].

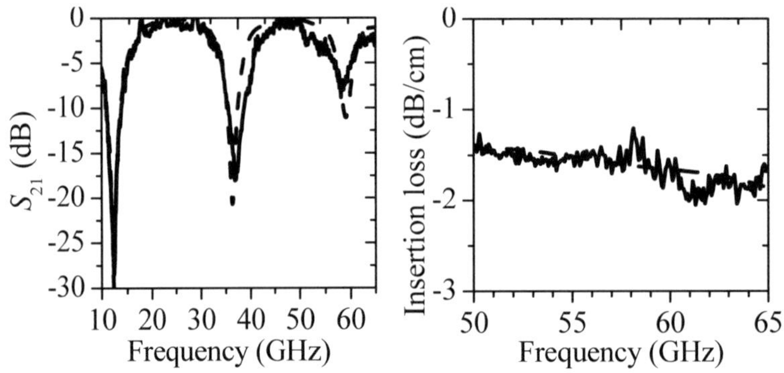

(a) (b)

Figure 19. a) Transmission coefficient S_{21} of the stub loaded microstrip line (l_s=4.58 mm, L=50 mm. (b) Insertion loss of a 50 Ω line. The numerical data assume ε_r=2.0 and tanδ=0.02. Measured (——) and computed (– –) data.

3.4.3. *Microstrip patch antenna*

The fabricated textile patch antenna operating at 60 GHz is shown in Fig. 20a. For measurement purpose, it is integrated with a V-connector. The flexibility of the antenna is demonstrated in Fig. 20b. The antenna was optimized to operate at 60 GHz using CST Microwave Studio. The reflection coefficient and radiation pattern of the textile antenna have been characterized in free space and on a parallepipedic skin-equivalent phantom (10×100×100 mm^3). The complex permittivity of the phantom equals that of human skin within the maximum error of 10% in the 57-64 GHz range [12].

First, the simulated and measured reflection coefficient is represented in Fig. 21. A frequency shift of only 2.5% is observed between computed and simulated results. It could be due to a change in the substrate permittivity and under-or over-etching of the microstrip line. Whereas at microwave frequencies patch antennas experience shift in resonance frequency [1], it can be seen that the reflection coefficient of the proposed antenna is immune from the human body proximity.

Measured radiation patterns in free space and on the homogeneous phantom were measured at 60 GHz. It was observed that the radiation pattern is very slightly affected by the phantom. The simulated and measured gains equal 4.3 dBi and 4.2 dBi, respectively. On the phantom, the maximum gain is decreased by 0.2 dB and 0.7dB in simulation and in measurement, respectively. Hence, whereas at microwaves patch antennas could be highly affected in terms of gain and efficiency [1], at millimeter waves the antenna performances remains unchanged.

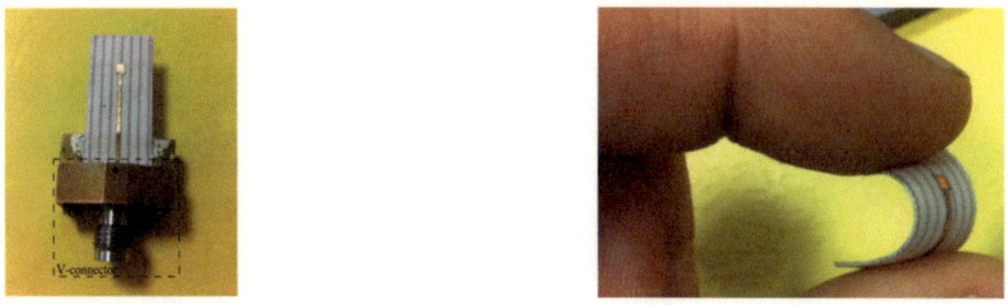

(a) (b)

Figure 20. Photography of the fabricated patch antenna with a V-connector.

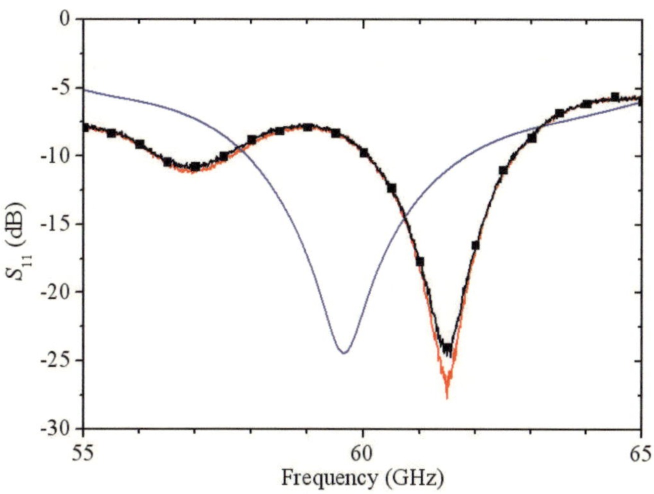

Figure 21. Reflection coefficient of the wearable patch antenna. —— Computed result in free space.—— Measured result in free space. —■—Measured result on the skin-equivalent phantom.

3.4.4. Microstrip patch antenna array

A microstrip-fed four-patch single-layer antenna array printed on the 0.2mm-thick textile has been designed (Fig. 22a) [29]. The array is fed by a 15.2mm-long microstrip line to avoid too strong reflections from the V-connector (Fig. 22). In practice, as textiles are more lossy than classical substrates, it is recommended to reduce the access line length as much as possible. Whereas the antenna could be fed using a central probe, (as shown in Fig. 22b), the microstrip feed line is the easiest solution to perform measurements on textile. We will discuss the impact of this microstrip line in terms of loss and distortion of the radiation pattern. The fabricated antenna integrated with a V-connector is shown in Fig. 23.

Its reflection coefficient S_{11} is measured using a 110-GHz Agilent 8510XF VNA and is shown in Fig. 24. Excellent agreement is obtained between simulated and measured results. The reproducibility of these results has been demonstrated and more information can be found in [29].

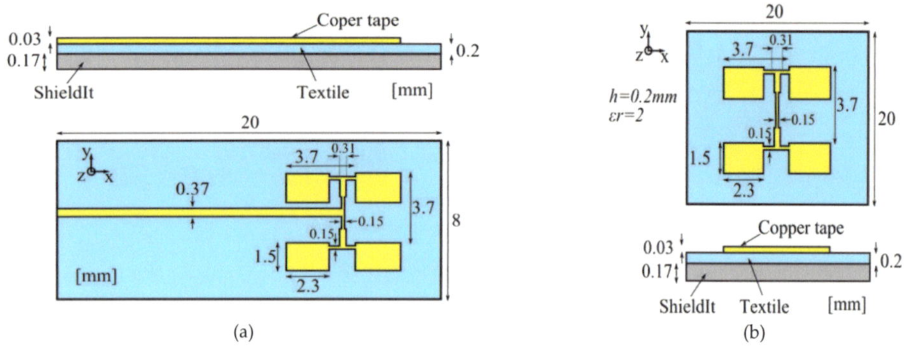

Figure 22. Layout of the microstrip antenna array printed on textile. (a) Antenna fed using a long microstrip line. (b) Antenna fed using a central coaxial probe.

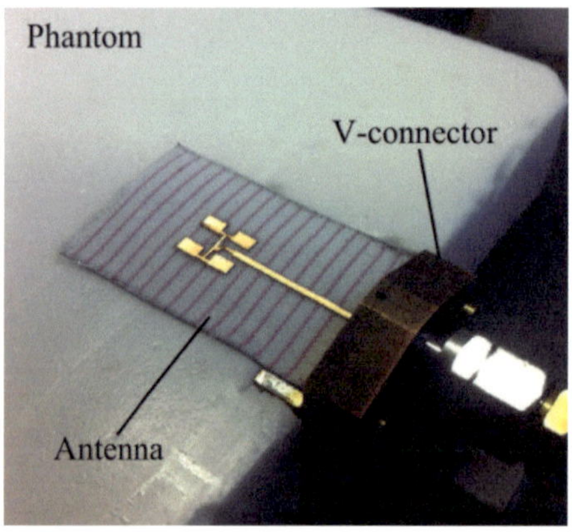

Figure 23. Measurement set-up on the skin-equivalent phantom for a distance between the ground plane and the phantom equal to *d*=0 mm.

The radiation patterns in E-and H-planes were measured in IETR's millimeter-wave anechoic chamber. The gain was measured by the comparison method with a 20-dBi standard horn, and the directivity is found from a 3D radiation pattern measurement. The co-polarization

http://dx.doi.org/10.5772/58816

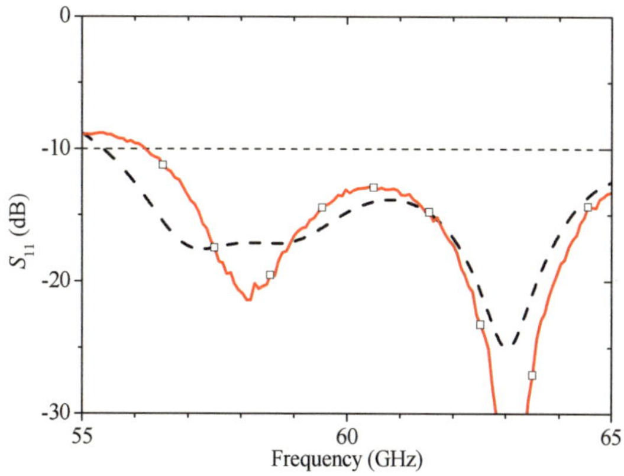

Figure 24. Reflection coefficient of the microstrip antenna array printed on textile. —□— Measured.---Simulated.

components measured in E-and H-planes at 60 GHz are in a good agreement with the computed ones (Fig. 25). In E-plane, the non-symmetry of the co-polarization component is attributed to the spurious radiation of feeding lines whose width is larger compared to standard substrates at millimeter waves like RT Duroid 5880 (see Section 3.3), or other commonly used substrate such as fused quartz or Alumina. The main characteristics of these three different substrates are compared in [29] with those of the textile used here. These data show that textile exhibits higher loss and that feeding lines are larger.

The simulated cross-polarization level remains lower than –20 dB at broadside in E-and H-planes. As expected and as already noticed in many previous papers (e.g. [24]), the measured values are much higher due to reflections and scattering on the V-connector and metallic support (Fig. 25b).

Besides, simulations have shown that the V-connector also affects the gain and directivity; therefore, for comparison purpose, these results are given for both configurations (i.e. with and without connector). The cross-polarization level could be further improved using a multilayer antenna design, e.g. [25]. However, as explained in Section 3.2, the latter is not recommended for on-body applications due to the relatively high SAR levels.

The effect of the central microstrip line exciting the antenna array has been investigated numerically comparing the radiation patterns of the proposed array (Fig. 22a) and those of a coaxial-fed array (Fig. 22b) [29]. These results (not shown here) demonstrate that the increase of the cross-polarization levels and side lobe levels in E-plane is due to the main feed line.

In addition, the gain, directivity and efficiency of these two antennas have been characterized (Table 4). High losses are experienced in the feed line (about 3.3 dB). In order to increase the antenna gain and efficiency, the feed line could be shortened or even suppressed (Fig. 22b).

Finally, the antenna performance (i.e. reflection coefficient and radiation) was tested after a number of hand washing cycles. The antenna was measured before and after washing when fully dried; its performance remained unchanged. However, to extend the life duration of the antenna, the authors would recommend waterproofing the whole antenna.

	Gain (dBi)		Directivity (dBi)		Efficiency (%)	
	Sim.	Meas.	Sim.	Meas.	Sim.	Meas.
Microstrip-fed array (Fig. 22a)	8.6	8.0	12.1	11.9	45	41
Coxial-fed array (Fig. 22b)	11.9	-	13.1	-	75	-

Table 4. Comparison of antenna performances in terms of gain, directivity and efficiency.

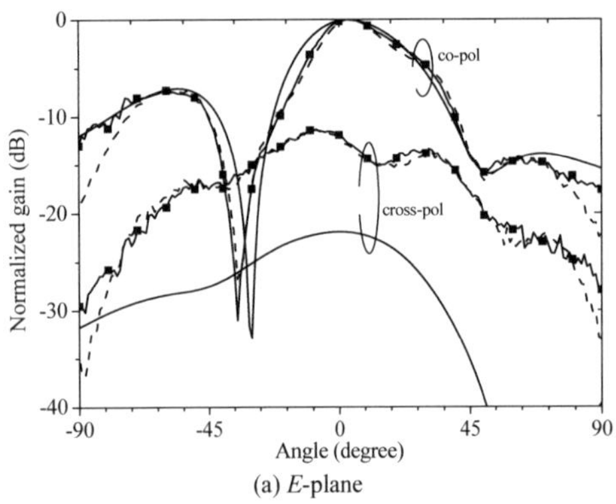

Figure 25. Normalized radiation patterns in co-and cross-polarization at 60 GHz. —— Simulation in free space. —■— Measurement in free space. --- Measurement on a skin-equivalent phantom.

3.5. Conclusion

Based on computed and measured results, antennas operating at millimeter-waves are very slightly sensitive to the human body. Besides, guidelines regarding the type of antennas, minimizing the interactions with the body, are provided. The feeding of the antenna is a critical point and aperture-coupled microstrip line-fed patch antennas should be avoided since it results in significantly higher body absorptions. A good alternative would be to use an aperture-coupled stripline-fed patch antenna instead.

Finally, textile antennas at millimeter-wave have been demonstrated with encouraging results. The textile can be accurately characterize and employed as antenna substrate. The textile antenna prototypes, fabricated using a simple and commercially compatible fabrication process, demonstrate excellent flexibility capabilities which would simplify the integration in clothes.

4. Antennas for on-body communications at millimeter waves

Whereas off-body communications appear to be a good solution at millimeter waves, on-body communications might be more challenging. In particular, significant shadowing effect from the human body is expected to make non-line-of-sight communications very difficult if not impossible. In [34], an on-body scenario has been numerically investigated in terms of propagation and demonstrates that short-range communications are achievable. The propagation issues are out of the scope of this Chapter; the readers can refer to the following papers for more details [33]-[35]. A few antennas optimized for on-body communications have been presented in the literature so far [31],[36],[37]. This Section will emphasize on the antenna performances in close proximity to the body.

4.1. Antenna requirements for on-body communications

On-body antennas should be as compact as possible to be integrated with a transceiver. As for off-body antennas, they have to be light weight and possibly conformable to the human body shape. Because of the high attenuation related to the propagation on a lossy dielectric (i.e. human body), medium-gain antennas (~12dBi) are required. The radiation pattern should be maximized toward the direction of propagation to minimize losses and make end-fire antennas excellent solutions. As the power is directed toward the body surface, absorptions inside the human is of uppermost concern.

4.2. End-fire antenna

A compact planar and flexible Yagi-Uda antenna covering the 57-64 GHz range designed for on-body communications is presented. The antenna is characterized in free space in terms of reflection coefficient, radiation pattern, and efficiency. The effect of the human body on the antenna characteristics is studied numerically and experimentally using a skin-equivalent phantom. The antenna performances are also studied under bending conditions. An on-body scenario is numerically investigated in terms of propagation.

4.2.1. Antenna model

High gain antenna is required for a line-of-sight path of human body dimensions. Furthermore, the maximum of the radiation pattern should be tangential to the body surface in order to reduce radiation off the body and thus minimizing interference among different BANs. Hence, a low-profile high-gain antenna with an end-fire radiation pattern printed on a 0.254mm-thick RT Duroid 5880 substrate (ε_r=2.2, $tan\delta$=0.003) is proposed. The layout is represented in Fig. 26. For measurement purpose the antenna prototype is mounted with a V-connector (Fig. 27).

4.2.2. Antenna performance in free space

The reflection coefficient S_{11} of the antenna array is measured with a 110 GHz vector network analyzer (Agilent 8510XF) using a V-connector (Fig. 27). The measured and simulated S_{11} (Fig. 28) are below-10 dB in the whole 57-64 GHz range. The numerical model does not include the V-connector.

The radiation patterns in *E*-and *H*-planes are plotted in Fig. 29. The simulated and measured radiation patterns at 60 GHz are in good agreement. The cross-polarization remains lower than-14 dB in the *E*-and *H*-planes at broadside. The gain was measured by the comparison

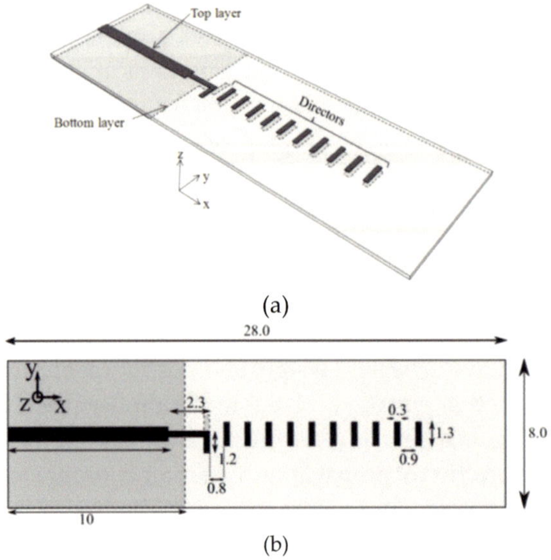

Figure 26. Layout of the printed Yagi-Uda antenna. Dimensions are in mm. (a) Three dimensions view. (b) Top layer.

Figure 27. Manufactured antenna with a V-connector.

method with a 20-dBi standard horn. At this frequency, the measured and computed gains equal 11.8 dBi and 12.1 dBi respectively. The losses of the V-connector (~0.8 dB) at 60 GHz were measured separately and taken out.

The antenna efficiency defined as the measured gain over the computed directivity equals to 86% at 60 GHz. It is in agreement with the simulated efficiency which equals 92%.

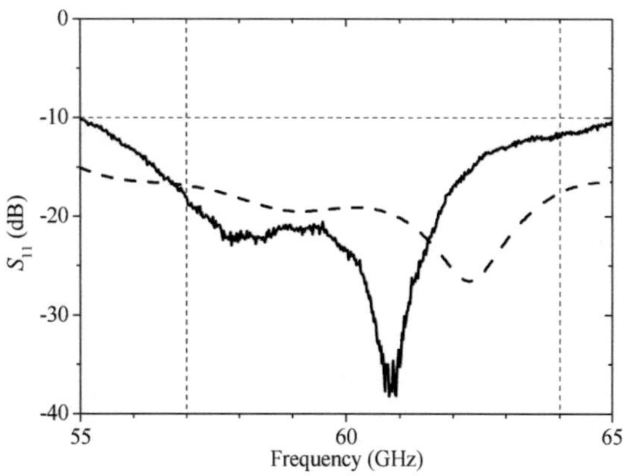

Figure 28. Measured and simulated reflection coefficient of the antenna in free space. --- Measurement. —— Simulation.

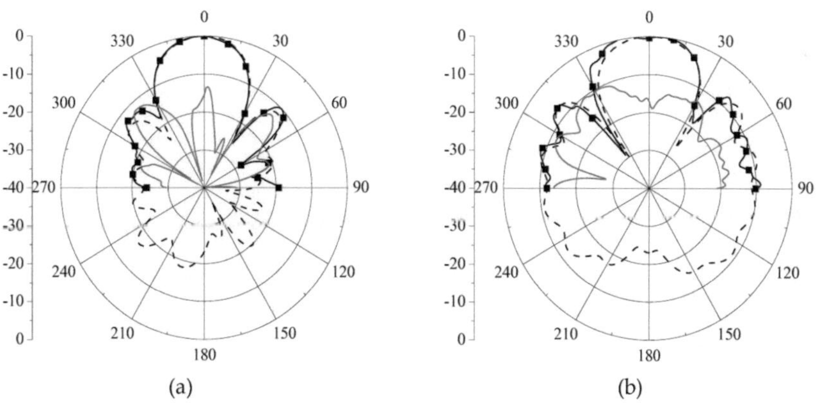

Figure 29. Measured and simulated radiation patterns in free space at 60 GHz in (a) *E*-and (b) *H*-planes. —■— Measured co-pol. —— Computed co-pol. —— Measured cross-pol.

4.2.3. *Antenna under bending conditions*

As it is difficult to keep the antenna flat in wearable applications, antenna performances under bending conditions is an important factor to be examined. The reflection coefficient and the *H*-plane radiation pattern are investigated when the antenna is placed on semi-cylindrical Rohacell HF51 foam with a radius *R* of 15mm (Fig. 30).

The chosen radius represents extremely severe test. The S_{11} is measured when the antenna is bent in the *H*-plane. The S_{11} remains below-10 dB in the whole 57-64 GHz range (Fig. 31).

Measured and simulated radiation pattern for *H*-plane bending are represented in Fig. 32. The maximum radiation follows the direction of the directors (-46°). Besides, the measured gain equals 11.1 dBi. This is in good agreement with the simulated gain (11.0 dBi). Compared to the gain in free space, a drop of 0.7 dB is observed in measurement.

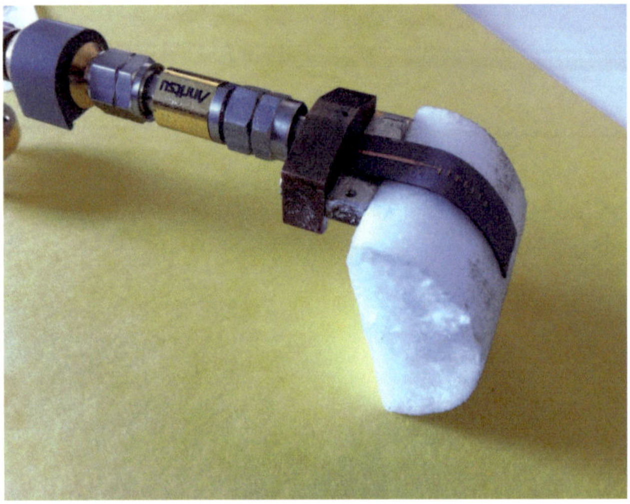

Figure 30. Bending antenna in the *H*-plane placed on a semi-cylindrical foam with *R*=15mm.

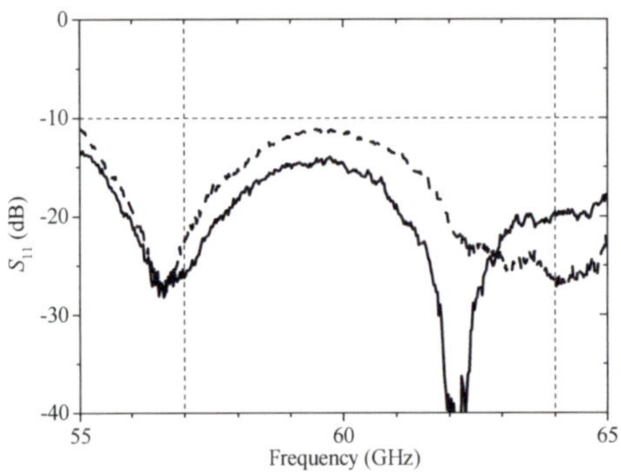

Figure 31. Measured reflection coefficient of the bent antenna mounted on semi-cylindrical foam. --- Flat. —— *R*=15mm.

4.2.4. *Antenna performances on the human body*

The antenna characteristics are assessed when placed on a skin-equivalent phantom (Fig. 33) in terms of reflection coefficient, radiation pattern, gain, and efficiency. The measured reflection coefficients of the antenna mounted on the phantom at different antenna/body spacing *h* are compared to that obtained in free space in Fig. 34. For h=5mm, the reflection coefficient is very slightly affected. For *h*=2mm, even though the S_{11} is much more affected and a frequency shift is observed, it remains below-10dB within the whole 57-64 GHz.

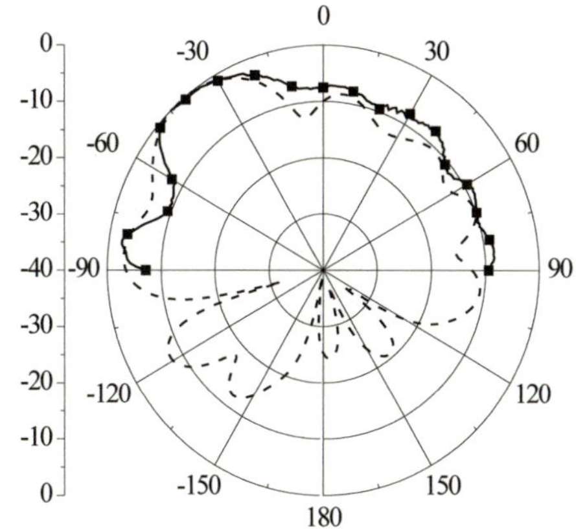

Figure 32. *H*-plane radiation pattern of the bent antenna (*R*=15mm) mounted on a semi-cylindrical foam. —■— Measured co-pol. ---- Computed co-pol.

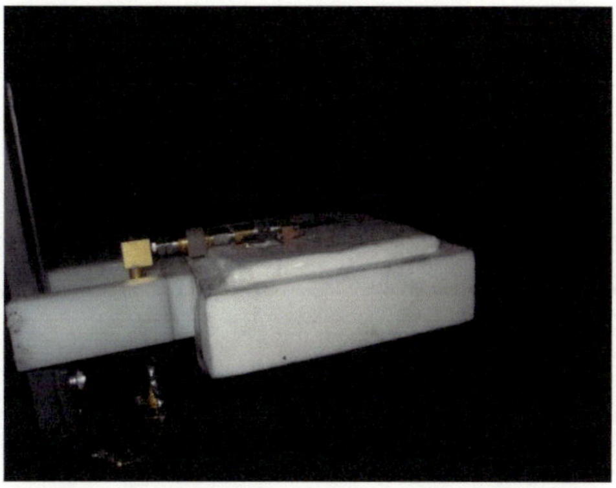

Figure 33. Antenna on the skin-equivalent phantom.

The measured radiation patterns in both *E*-and *H*-planes at 60 GHz of the antenna placed on the skin-equivalent phantom are represented in Fig. 35 for *h*=5.6mm and *h*=2mm. Both *E*-and *H*-planes are strongly affected by the human body because of reflection on and absorption in the body.

Here, the radiation pattern is titled because of reflections occurring at the air/phantom interface. A tilt of 10° and 21° is observed for an antenna/body spacing of 5.6mm and 2mm, respectively. The simulated and measured gains and the simulated efficiency are summarized in Table 5 for different antenna/body spacing. The efficiency decreases with *h*. However, the maximum gain of the antenna increases on the phantom (up to 3dB increase for *h*=5.6mm). Compared to the free space configuration, radiations toward the human body are significantly reduced because of reflections from and absorptions in the human body. Hence, when the antenna is mounted on the phantom, its performance remains satisfactory in terms of reflection coefficient, radiation pattern, and efficiency.

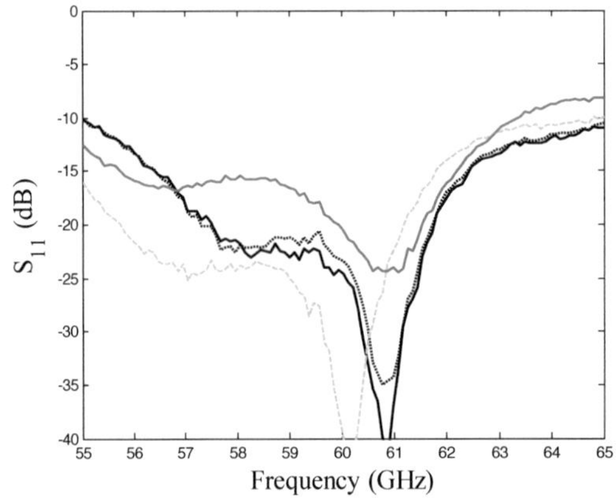

Figure 34. Measured reflection coefficient of the antenna array on the homogeneous phantom. — Free space. ▪▪▪▪▪ On phantom with *h*=5.6mm.---On phantom with *h*=2mm. — On phantom with *h*=0mm.

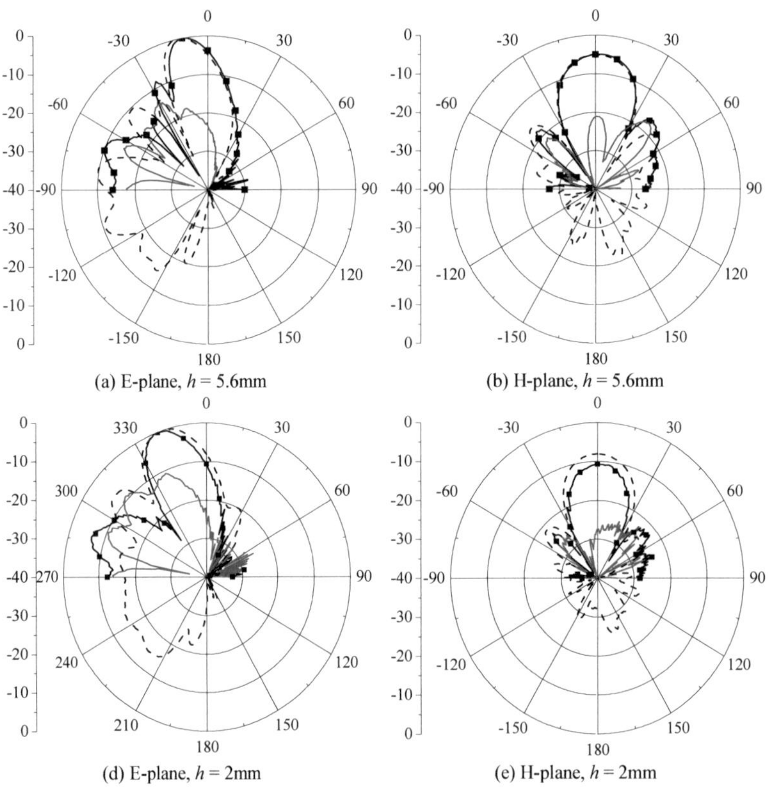

Figure 35. Measured and simulated radiation patterns on the skin-equivalent phantom at 60 GHz. —■— Measured co-pol. —— Computed co-pol. —— Measured cross-pol.

Antenna/phantom separation h (mm)	Gain (dBi)		Simulated efficiency (%)
	Simulated	Measured	
∞	12.1	11.8	92.2
5.6	15.1	15.2	74.9
2	13.6	13.6	68.2

Table 5. Antenna gain and efficiency for different antenna/body spacing.

4.3. Conclusions

A compact planar Yagi-Uda antenna covering the 57-64 GHz range has been designed for on-body communications. The effect of the human body on the antenna characteristics has been studied numerically and experimentally using a skin-equivalent phantom. It was shown that the distance between the antenna and the human body has a strong impact on the antenna performances. The antenna was also studied under bending conditions demonstrating satisfactory performances. The same antenna has been successfully optimized and fabricated on textile [31].

5. Conclusion

Challenges and progress in antennas and their interaction with the human body in body-centric scenarios at millimeter-wave frequencies have been presented in this Chapter. Recent progress in manufacturing and modeling experimental phantoms has been discussed. These phantoms play a key role in characterizing the antenna performance in close proximity to the human body.

As far as off-body communications are concerned, it was shown that the feeding type is an important factor since it can strongly influence absorption in the human body. In addition, performances of patch antenna arrays in close proximity to the human body have been evaluated showing very slight impact on the antenna performance. Besides, a textile patch antenna array, operating at millimeter waves, was successfully demonstrated using a commercially-available textile. An accurate and low-cost fabrication process has been introduced. Research work should now be focused on the interconnections between textile antennas and Radio Frequency Integrated Circuits (RFIC) since this issue has not been tackled yet.

Finally, as end-fire antennas appear to be the best solution for on-body communications, a Yagi-Uda antenna has been investigated. It appears that the antenna radiation pattern is strongly affected by the separation between the antenna and the human body. This antenna is robust against bending which is an important asset if this antenna would be implemented on textile as shown in [31]. Other antenna designs for on-body communications were introduced in [38].

While these results are promising, millimeter-wave wireless systems still have considerable challenges to overcome to enable mass commercialization. First, mm-wave wireless must address challenging RF impairments such as fast fading and delay spread conditions making demodulation and equalization particularly difficult with reasonable architectures and complexities. Second, millimeter-wave transceivers require giga-samples per second (GS/s) scale data-converters with considerable resolutions leading to high power consumption (even in advanced technology nodes). Finally, mm-wave schemes must prove themselves competitive with advanced and adaptive modulation and channel coding schemes (256 QAM and beyond) like 802.11ac 5th generation WiFi that can also reach high data rates (6.77 Gbit/s nominal) while being built upon existing wireless hardware and infrastructure in the 5.83 GHz ISM band.

Acknowledgements

This work was supported by French National Research Agency (ANR) under Grant ANR-09-RPDOC-003-01 (Bio-CEM project), by Labex CominLabs (ANR program "Investing for the Future" ANR-10-LABX-07-01) and Brittany Region under ResCor/BoWi project and by National Center for Scientific Research (CNRS), France. Part of this work was performed using HPC resources from GENCI-IDRIS (grant 2013-050779).

Author details

Nacer Chahat[1], Maxim Zhadobov[2] and Ronan Sauleau[2]

1 NASA Jet Propulsion Laboratory/California Institute of Technology, Pasadena, CA, USA

2 University of Rennes 1, Rennes, France

References

[1] P. S. Hall and Y. Hao, "Antennas and propagation for body centric communications systems," Artech House, Norwood, MA, 2006, ISBN-10: 1-58053-493-7.

[2] D. Guha and Y. M. M. Antar, "Microstrip and printed antennas: new trends, techniques and applications," Wiley-Blackwell, 2011, ISBN-10: 0470681926.

[3] T. Baykas, C. S. Sum, Z. Lan, J. Wang, M. A. Rahman, and H. Harada, "IEEE 802.15.3c: the first IEEE wireless standard for data rates over 1 Gb/s," *IEEE Communications Mag.*, vol. 49, no. 7, pp. 114–121, Jul. 2011.

[4] S. L. Cotton, W. G. Scanlon, and P. S. Hall, "A simulated study of co-channel inter-BAN interference at 2.45 GHz and 60 GHz," *Europ. Wirel. Techn. Conf.*, Paris, France, pp. 61–64, Sep. 2010.

[5] M. Kojima, et al., "Acute ocular injuries caused by 60-Ghz millimeterwave exposure," *Health Phys.*, vol. 97, no. 3, pp. 212–218, Sept. 2009.

[6] H. A. Kues, S. A. D'Anna, R. Osiander, W. R. Green, and J. C. Monahan, "Absence of ocular effects after either single or repeated exposure to 10mW/cm2 from a 60GHz CW source," *Bioelectromagnetics*, vol. 20, no. 8, pp. 463-473, Dec. 1999.

[7] F. A. Duck, "Physical Properties of Tissue: a comprehensive reference network," U.K.: Academic, ISBN 0122228006, 1990.

[8] O. P. Gandhi and A. Riazi, "Absorption of millimeter waves by human beings and its biological implications," *IEEE Trans. Microwave. Theory Tech.*, vol. 34, no. 2, pp. 228–235, Feb. 1986.

[9] C. M. Alabaster, "Permittivity of human skin in millimetre wave band," *Elec. Lett.*, vol. 39, no. 21, pp. 1521–1522, Oct. 2003.

[10] S. Gabriel, R. W. Lau, and C. Gabriel, "The dielectric properties of biological tissues: III. Parametric models for the dielectric spectrum of tissues," *Phys. Med. Biol.*, vol. 41, no. 11, pp. 2271–2293, Nov. 1996.

[11] S. I. Alekseev and M. C. Ziskin, "Human skin permittivity determined by millimeter wave reflection measurements," *Bioelectromagnetics*, vol. 28, no. 5, pp. 331–339, Jul. 2007.

[12] N. Chahat, M. Zhadobov, and R. Sauleau, "Broadband tissue-equivalent phantom for BAN applications at millimeter waves," *IEEE Transactions on Microwave Theory and Techniques*, vol. 60, no. 7, pp. 2259–2266, Jul. 2012.

[13] N. Chahat, M. Zhadobov, R. Sauleau, and S. Alekseev, "New method for determining dielectric properties of skin and phantoms at millimeter waves based on heating kinetics," *IEEE Transactions on Microwave Theory and Techniques*, vol. 60, no. 3, pp. 827–832, March 2011.

[14] P. F. M. Smulders, "Impact of regulations on feasible distance between 60 GHz devices," *Europ. Conf. Antennas Propag.*, EuCAP'2010, Barcelona, Spain, Apr. 12–16, 2010.

[15] S. L. Cotton, W. G. Scanlon, and B. K. Madahar, "Millimeter-wave soldier-to-soldier communications for covert battlefield operations," *IEEE Communications Mag.*, vol. 47, no. 10, pp. 72–81, Oct. 2009.

[16] P. F. M. Smulders, "Statistical characterization of 60-GHz indoor radio channels," *IEEE Trans. Antennas Propag.*, vol. 57, no. 10, pp. 2820–2829, Oct. 2009.

[17] R. C. Daniels, J. N. Murdock, T. S. Rappaport, and R. W. Heath, "60 GHz wireless: up close and personal," *IEEE Microwave Magazine,* vol. 11, no. 7, pp. 44–50, Dec. 2010.

[18] S. Collonge, G. Zaharia, and G. El Zein, "Wideband and dynamic characterization of the 60 GHz indoor radio propagation—future home WLAN architectures," *Ann. Telecommun.*, vol. 58, no. 3–4, Mar.–Apr. 2003.

[19] K. L. Wong and C. I. Lin, "Characteristics of a 2.4-GHz compact shorted patch antenna in close proximity to a lossy medium," *Microw. Opt. Technol. Lett.*, vol. 45, no. 6, pp. 480–483, Jun. 2005.

[20] L. Vallozzi, W. Vandendriessche, L. Vallozzi, W. Vandendriessche, H. Rogier, C. Hertleer, and M. L. Scarpello, "Wearable textile GPS antenna for integration in protective garments," *Europ. Conf. Antennas Propag.*, EuCAP'2010, Barcelona, Spain, Apr. 12–16, 2010.

[21] M. M. Khan, A. Alomainy, and Y. Hao, "Dual band and diverse radiation pattern antenna for power efficient and reliable on-body and off-body communications for healthcare applications," *IEEE Ant. Propag. Society Int. Symp.*, Spokane (WA), Jul. 2011.

[22] "Guidelines for limiting exposure to time-varying electric, magnetic, and electromagnetic fields (up to 300 GHz)," *Health Phys.*, vol. 74, no. 4, pp. 494–522, 1998.

[23] M. F. Karim, Y.-X. Guo, M. Sun, J. Brinkhoff, L. C. Ong, K. Kang, and F. Lin, "Integration of SiP-Based 60-GHz 4 × 4 Antenna Array With CMOS OOK Transmitter and LNA," *IEEE Trans. Microw. Theory Tech.*, vol. 59, no. 7, pp. 1869–1878, Jul. 2011.

[24] N. Chahat, M. Zhadobov, L. Le Coq, S. Alekseev, and R. Sauleau, "Characterization of the Interactions between a 60-GHz Antenna and the Human Body in an Off-Body Scenario," *IEEE Transactions on Antennas and Propagation*, vol. 60, no. 12, pp. 5958–5965, Dec. 2012.

[25] P. Coquet, R. Sauleau, K. Shinohara, and T. Matsui, "Multi-layer microstrip antennas on quartz substrates. Technological considerations and performance at 60 GHz," *Microw. Opt. Technol. Lett.*, vol. 40, no. 1, pp. 41–47, Jan. 2004.

[26] B. Biglarbegian, M. Fakharzadeh, D. Busuioc, M. R. Nezhad-Ahmadi, and S. Safavi-Naeini, "Optimized microstrip antenna arrays for emerging millimeter-wave wireless applications," *IEEE Trans. Antennas Propag.*, vol. 59, no. 5, pp. 1742–1747, May 2011.

[27] B. Chantraine-Bares, R. Sauleau, L. Le Coq, and K. Mahdjoubi, "A new accurate design method for millimeter-wave homogeneous dielectric substrate lens antennas of arbitrary shape," *IEEE Trans. Antennas Propag.*, vol. 53, no. 3, pp. 1069–1082, March 2005.

[28] M. Zhadobov, N. Chahat, R. Sauleau, C. Le Quement, Y. Le Dréan, "Millimeter-wave interactions with the human body : state of knowledge and recent advances," *International Journal of Microwave and Wireless Technologies*, vol. 3, no. 2, pp. 237-247, April 2011.

[29] N. Chahat, M. Zhadobov, and R. Sauleau, "60-GHz textile antenna array for body-centric communications," *IEEE Transactions on Antennas and Propagation*, vol. 61, no. 4, pp. 1816–1824, Apr. 2013.

[30] S. Hage-Ali, N. Tiercelin, P. Coquet, R. Sauleau, H. Fujita, V. Preobrazhensky, and P. Pernod, "A millimeter-wave microstrip antenna array on ultra-flexible micromachined polydimethylsiloxane (PDMS) polymer," *IEEE Antennas Wirel. Propag. Lett.*, vol. 8, pp. 1306–1309, 2009.

[31] N. Chahat, M. Zhadobov, and R. Sauleau, "Wearable end-fire textile antenna for on-body communications at 60 GHz," *IEEE Antennas Wirel. Propag. Letters*, vol. 11, pp. 799–802, 2012.

[32] I. Locher, M. Klemm, T. Kirstein, and G. Troster, "Design and characterization of purely textile patch antennas," *IEEE Trans. Advanced Packaging*, vol. 29, no. 4, pp.777–788, Nov. 2006.

[33] T. Mavridis, L. Petrillo, J. Sarrazin, D. Lautru, A. Benlarbi-Delai, and P. De Doncker, "Theoretical and Experimental Investigation of a 60 GHz Off-Body Propagation Model," *IEEE Transactions on Antennas and Propagation*, in press.

[34] N. Chahat, G. Valerio, M. Zhadobov, and R. Sauleau, "On-body propagation at 60 GHz," *IEEE Transactions on Antennas and Propagation*, vol. 61, no. 4, pp. 1876–1888, Apr. 2013.

[35] A. Guraliuc, M. Zhadobov, V. Guido, N. Chahat, and R. Sauleau, "Effect of Textile on the Propagation Along the Body at 60 GHz," *IEEE Transactions on Antennas and Propagation*, in press.

[36] X. Y. Wu and P. S. Hall, "Substrate integrated wavegide Yagi-Uda antenna," *Electron. Lett.*, vol. 11, no. 23, pp. 1541–1542, Nov. 2010.

[37] X. Y. Wu, L. Akhoondzadeh-Asl, and P. S. Hall, "Printed Yagi-Uda array for on-body communication channels at 60 GHz," *Microw. Opt. Technol. Lett.*, vol. 53, no. 12, pp. 2728–2730, Dec. 2011.

[38] A. Pellegrini, A. Brizzi, L. Zhang, K. Ali, Y. Hao, X. Wu, C. C. Constantinou, Y. Nechayev, P. S. Hall, N. Chahat, M. Zhadobov, and R. Sauleau, "Antennas and propagation for body centric wireless communications at millimeter wave frequencies: a review," *IEEE* Antennas and Propagation Magazine, vol. 55, no. 4, Nov. 2013.

Chapter 3

Low Cost Compact Multiband Printed Monopole Antennas and Arrays for Wireless Communications

Qi Luo, Jose Rocha Pereira and Henrique Salgado

Additional information is available at the end of the chapter

http://dx.doi.org/10.5772/58815

1. Introduction

Compact size printed multiband monopoles are of interest for a variety of applications such as WLAN, RFID and mobile terminals. If the antennas can be fabricated with a planar structure using PCB techniques, the cost can be kept low and the fabrication process is greatly simplified.

It is well-known that the size reduction will decrease the radiation efficiency of the antenna especially when its size is very small compared to the free space wavelength at its lowest resonant frequency. Conventional high permittivity substrates can be employed to reduce the size of the microstrip antenna (e.g. printed microstrip patch) but raises other design issues. When using this approach the bandwidth of the antenna is decreased and the surface wave propagations are excited, which can lead to the scan blindness if a beam-steerable phased array is built based on this antenna element. Therefore, it is important to investigate techniques for the design of compact and low cost microstrip antennas with promising radiation characteristics.

This chapter discusses various techniques of designing low cost, small-size printed monopole antennas and it is organized as follows. In section 2, antenna miniaturization techniques, based on fractal geometries and the use of lumped elements into the radiating element are discussed. In section 3, a low cost multiband printed planar monopole for mobile terminals is presented. This printed monopole exhibits five resonant frequencies and covers the desired frequency bands for mobile, WiMax and WLAN operations. Then in section 4, the design of a small size printed monopole array is addressed and two examples are given, one of which can be employed to increase the gain of the antenna and the other is suitable for MIMO applications of portable devices. Finally, recent developments in the field of low cost compact printed monopoles and arrays are discussed in section 5.

2. Compact and electrically small printed multiband monopoles

The currently available and future commercial wireless systems require antennas having wide bandwidth to support higher data rate and be able to operate at multiple frequency bands defined by various protocols. Compact size printed monopole antennas are important for the wireless applications due to its advantages of easy fabrication, omnidirectional radiation and wide operation bandwidth.

It is known that to reach the resonant condition, the dimension of the antenna must be a fraction of the wavelength at its resonant frequency. This means that the lower resonant frequency is, the larger the size of the antenna will be. From the limitations of the electrically small antennas defined by Chu [1], it is clear that antennas of smaller size always exhibit a higher quality factor whereas the bandwidth of an antenna is inversely proportional to the quality factor. As such the size reduction of an antenna will lead to the deterioration of its radiation performance. Therefore, compact multiband antenna design with promising radiation performance has attracted much research interests.

To be a low cost solution, it is desirable to fabricate the monopoles in PCB technology using only a single layer of substrate and a planar structure. Being a planar structure, the radiating element must have a geometry that can excite higher modes within a limited volume, to have a multiband operation. One approach that can be employed is to use fractal geometries to design compact multiband printed monopole antennas. Fractal geometry is a family of geometries that have the characteristics of inherent self-similar or self-affinity, which were used to describe and model complex shapes found in nature such as mountain ranges, waves and trees [2]. Recently, fractal techniques have been brought to the field of electromagnetic theory, a research field which has been called fractal electrodynamics; it has also been implemented in antenna design and named "fractal antenna engineering". This topic has been attracting much research interest. There are several advantages of using fractal geometries. First of all, it can reduce the size of the antenna, which makes it a good candidate for antenna miniaturization. Fractal geometries are self-filling structures that can be scaled without increasing the overall size. This characteristic provides opportunities for antenna designers to explore new geometries suitable for small antenna design. Secondly, fractal is a geometry that is self-repeated at different scales, which means that the fractal technique can be explored for designing antenna with multiple band operation and similar radiation patterns.

It is important to point out that although fractal geometries are self-filling structures that can be scaled without increasing the overall size, not all the geometries can contribute to the compact antenna design. Previous research found that some fractal geometries such as Hilbert and Peano curves, which exhibit a high degree of space filling, cannot effectively reduce the resonant frequency of the antenna due to the cancelling of the current between closely spaced lines [3].

http://dx.doi.org/10.5772/58815

2.1. Dual band fractal monopole

Studies show that Minkowski Island geometry is a good candidate for the design of multiband printed monopole antennas. Compared to other fractal geometries such as Hilbert curves, Minkowski Island geometry can work more efficiently with respect to the frequency reduction, due to its meandered-like configuration [3]. As demonstrated in Figure 1, when a Hilbert curve is employed to design a printed monopole antenna, the closely spaced lines can cause a large amount of current cancellation compared to the Minkowski Island geometry, which means that the effective electrical length of the Hilbert Curve antenna cannot benefit much from using such space filling geometry.

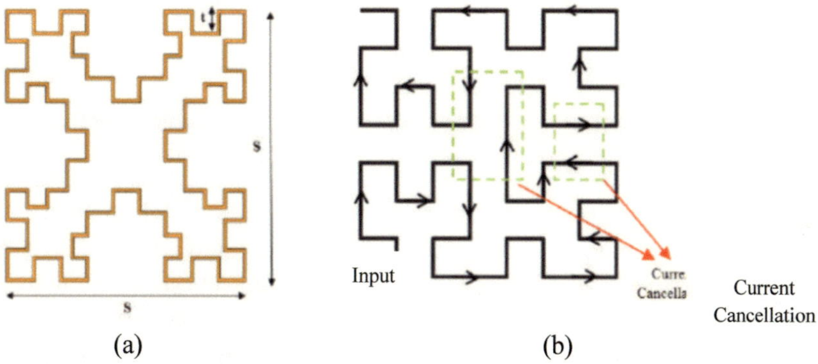

Figure 1. (a) The 2nd iteration of Minkowski Island geometry; (b) current cancellation of a Hilbert Curve (b) [3].

In [4], two compact dual-band printed fractal monopoles for WLAN applications were presented. These two monopoles are designed using the 1st and 2nd iteration of the Minkowski Island, as presented in Figure 2. Figure 2(a) shows the printed monopole based on the 1st iteration Minkowski Island. Its size is 28 mm ×18 mm with a partial ground plane having a width of 35 mm and length of 10 mm on the back side of the substrate. The width of the microstrip line is 0.5 mm. Figure 2(b) shows another proposed fractal monopole using the 2nd iteration Minkowski Island. Its size is 21.5 mm ×18 mm and the size of the ground plane is 30 mm ×10 mm. The depth *t*, shown in Figure 1, is 1/4 of the side length (*s*/4) at each iteration for both antennas. The line widths of both antennas were set based on two factors: the antenna input impedance and the fact that the microstrip line needs to be narrow enough to avoid the intersection between adjacent lines. This issue is more significant for fractal of higher iterations. As a result of using a higher iteration fractal, narrower microstrip line needs to be used and the width of the microstrip line is reduced to 0.25 mm. Both of the proposed antennas are printed on the top side of the substrate, 0.813 mm thick Roger 4003 with relative permittivity ε_r=3.38, while the ground plane is printed at the bottom side. Behind the antenna elements, there is no ground plane.

Both of the proposed monopole antennas have a compact size compared to conventional printed monopole antennas, which need to have a length of approximately a quarter of

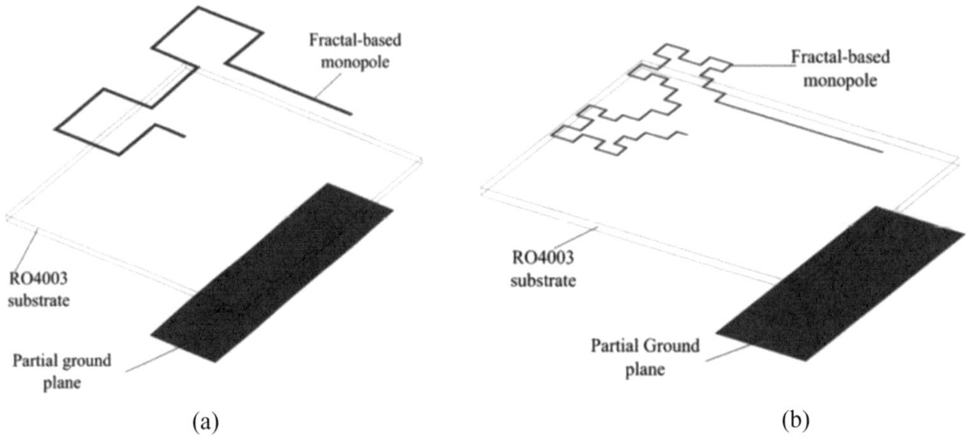

Figure 2. Exploded view of the fractal monopole antenna with geometry of: (a) 1st iteration Minkowski Island; (b) 2nd iteration of Minkowski Island proposed in [4].

wavelength. Moreover, it is found that without using any additional impedance matching techniques, both of these two proposed antennas exhibit good impedance match at multiple frequency bands, which is confirmed by the measurement results shown in Figure 3. These results show that the printed monopole of 1st iteration of Minkowski Island exhibits 10-dB return loss from 2.30-2.48 GHz, 3.3-3.7 GHz and 4.9-6.0 GHz, which covers the entire required frequency bands for 802.11a/b/g and WiMAX communications. For the 2nd iteration of Minkowski Island fractal monopole, the 10-dB return loss bands are 2.31-2.47 GHz and 5.0-5.5 GHz, which covers the two desired frequency bands for WLAN 802.11b/g standards.

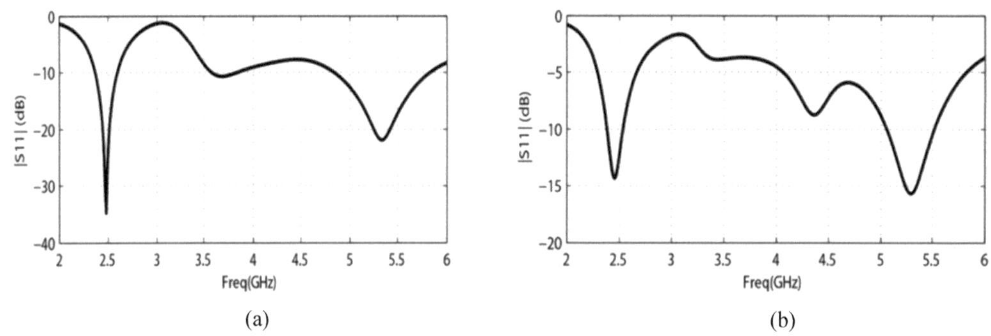

Figure 3. Measured reflection coefficient of the proposed antenna with the: (a) 1st iteration and (b) 2nd iteration of Minkowski Island geometry [4].

The measured results also show that the radiation patterns at the H-plane are almost isotropic and in the E-plane they exhibit broadside radiation patterns, as expected. The measured maximum gain is around 1.5 dB at 2.45 GHz and 2.3 dB at 5.2 GHz for both antennas. According to the simulation results, the radiation efficiency is 94% and 88% at 2.45 GHz, 97% and 93% at 5.26 GHz for the printed monopole of 1st and 2nd iteration of Minkowski Island, respectively. From these results, it is found that although higher size reduction can be achieved using higher iteration of the fractal geometry, the bandwidth as well as the radiation efficiency also decreases. This should be considered as a trade-off between size reduction and antenna performance. Figure 4 shows the measured E-and H-plane radiation pattern of the printed monopole with 2nd iteration of Minkowski Island at its dual resonant frequencies. Measurements showed that both antennas have similar radiation patterns so the measurement results for the other prototypes (the printed monopole of 1st iteration of Minkowski Island) are not given to avoid repetition.

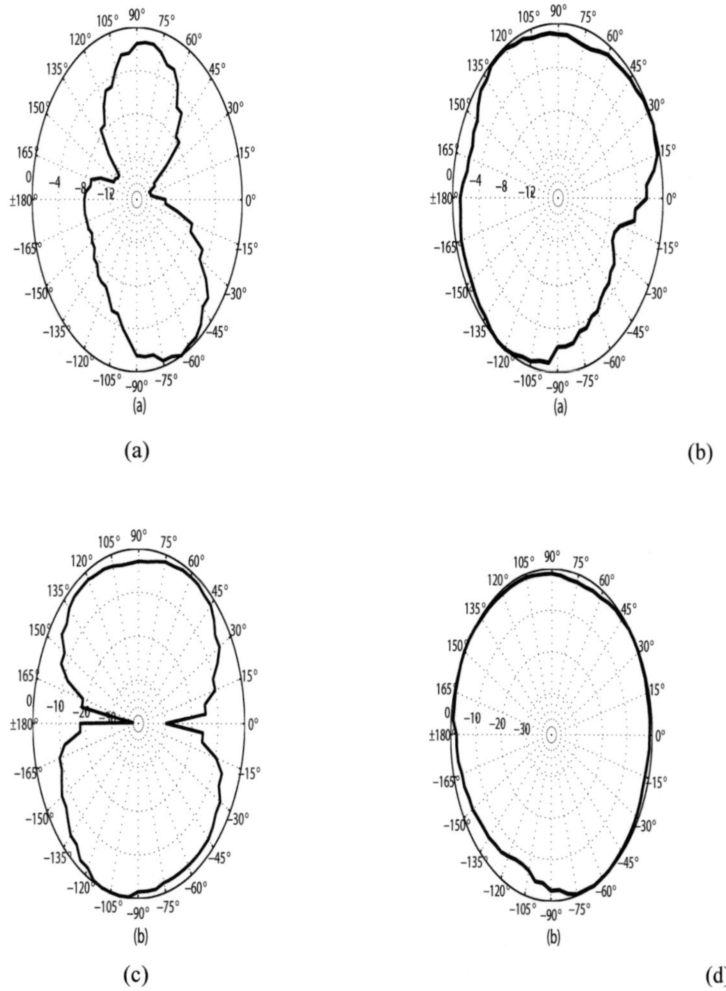

Figure 4. Measured radiation patterns of the proposed antenna with the 2nd iteration of Minkowski Island: (a) at 2.45°GHz, E-plane; and (b) at 2.45°GHz, H-plane; (c) at 5.3°GHz, E-plane and (d) at 5.3°GHz, H-plane [4].

2.2. Fractal ILA antenna

The frequency ratio of the multiband fractal antenna is investigated in [4]. It is shown that as a multiband antenna, the frequency ratio of the fractal antenna using the Minkowski geometry is nearly fixed. This indicates that in order to extend the fractal technique to other multiband antennas design, there is a need to explore an effective solution to overcome this limit. One technique that can be employed to extend the frequency ratio of the fractal-based multiband antenna design is to combine the fractal geometry with the meander line. One compact antenna suitable for a commercial wireless USB device is proposed by using such technique [5]. Since the objective is to design a printed fractal monopole antenna for WLAN USB dongle applications, based on the industrial requirement, the overall size of this antenna including the ground plane is chosen to be 20 mm×60 mm and the available space for antenna design is limited to no more than 20 mm × 10 mm.

A variation of the Koch fractal, which also can be referred as Cohen dipole fractal geometry, was used in this design. The Cohen dipole geometry, which is a variation of Koch fractal, was first proposed by Nathan Cohen [6] to design a dipole antenna with the feeding at the center position. Different from a conventional printed monopole antenna, the antenna radiating element is printed on the same layer of the ground plane. This type of antenna is named Inverted-L Antenna (ILA). As a typical printed monopole antenna, the antenna element with the feeding line and the ground plane are printed at the top and bottom side of the substrate, respectively. Meanwhile, the feeding port is located at the end of the substrate as shown in Figure 5. This might be a problem in a practical industrial design as other components, such as RF module, also need to be mounted on the same ground plane. Such problem can be solved by using ILA antenna.

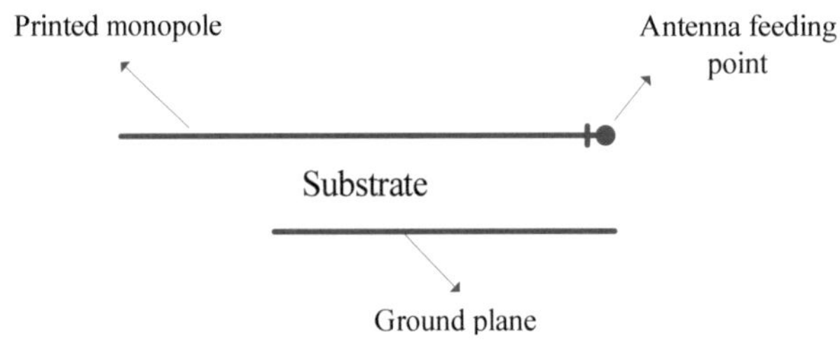

Figure 5. A typical configuration of the conventional printed monopole antenna

Figure 6 shows the exploded view of the proposed printed fractal ILA antenna in [5]. This antenna is designed on the Roger 4003 substrate with dielectric constant of 3.38 and thickness of 0.813 mm. The space occupied by the monopole antenna on the substrate is 10 mm ×20 mm and the size of the ground plane is 50 mm×20 mm, which is a typical size for a USB dongle. This structure was further optimized by doing numerical simulations in Ansoft HFSS to achieve a better impedance match at the required frequency bands. It is found that the size of the fractal geometry is critical in defining both the resonant frequencies while the existence of the horizontal microstrip line plays the role of adjusting the resonant frequencies to the desired region. Without the horizontal microstrip line, it is found that the proposed antenna can only exhibit resonances at around 2 and 6 GHz, which fails to cover the desired frequencies for

WLAN dual-band applications. However, adding the horizontal microstrip line with the appropriate length, the frequency ratio of the fractal antenna can be more controlled. After further optimization of the reflection coefficient over the desired frequency band, the width of the vertical microstrip line was chosen to be 1 mm and the width of the horizontal microstrip line was 0.5 mm. For the fractal, the width of the microstrip line was set to 0.35 mm.

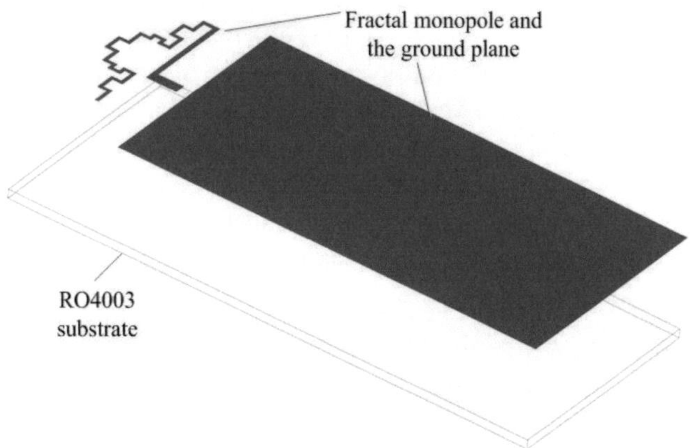

Figure 6. The exploded view of the printed fractal ILA antenna proposed in [5]

Figure 7 compares the reflection coefficient between simulation and measurement results. It can be seen that there is a good agreement between the simulated and measured reflection coefficient. The experimental result indicates that the proposed antenna has a S11<-10 dB with a bandwidth from 2.25 to 2.60 GHz and 5.06 to 5.62 GHz.

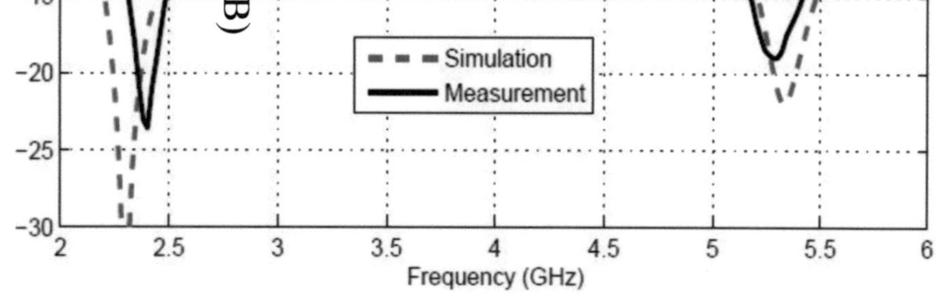

Figure 7. Comparison of the simulated and measured S11 of the fractal ILA in [5]

2.3. Compact Printed Monopole Antenna with Chip Inductor

Besides using fractal techniques to design printed monopoles of reduced size, another technique which can be employed is to introduce a lumped element, more specifically a chip inductor, into the antenna radiating element. In this way, the effective electrical length of the printed monopole is increased by the actual chip inductor instead of employing the fractal geometries that can bend a microstrip line of large length in a finite area.

Figure 8 shows the layout of the printed monopole antenna proposed in [7]. This antenna has a two-armed structure and for such monopole antenna the resonant frequency can be created by letting the overall length of each arm to be approximately a quarter of its effective wavelength on the substrate. The chip inductor is embedded in the middle of the left arm and generally speaking, the higher the value of the inductance, the lower the resonant frequency that can be achieved. However, increasing the inductance will also reduce the bandwidth and radiation efficiency of the antenna, which is the reason why a chip inductor with a higher inductance is not chosen in this study.

This antenna was printed on a Roger 4003 substrate with relative permittivity of 3.38 and thickness of 0.813 mm. The antenna and ground plane were printed on different sides of the substrate and there is no copper below the antenna section. The area of this antenna is only 10 mm ×10.5 mm, which is only 0.08 $\lambda_{2.4GHz}$ ×0.084 $\lambda_{2.4GHz}$, where $\lambda_{2.4GHz}$ represents the free space wavelength at 2.4 GHz. The higher band of the antenna is determined by the overall length L4+L5, which is approximately a quarter of a wavelength at 5.3 GHz. With the chip inductor, the overall length L1+L2+L3, which determines the lower band resonant frequency, is only 12.5 mm. This value is smaller than the length required for conventional monopole antennas. After adding the chip inductor, the resonant frequencies of the lower and higher band can be tuned by respectively changing the length of the arm L3 and L5, as demonstrated in the next section. By optimizing the length and width of each arm, this antenna is tuned to resonate at the desired frequencies. For more details about how to choose the value of the chip inductor and model it in the EM simulation software, readers can refer to [7].

The measured radiation patterns of the two-armed monopole antenna at 2.45 and 5.3 GHz show that the antenna has omnidirectional radiation patterns as a typical monopole antenna. Table 1 presents both measured and simulated results, as well as the calculated quality factor of the proposed monopole antenna. The quality factor was calculated based on the equations given below, taken from [8]:

$$Q(\omega_0) = \frac{2\sqrt{\beta}}{FBWv(\omega_0)}$$

$$FBWv(\omega_0) = \frac{\omega_+ - \omega_-}{\omega_0}$$

$$\sqrt{\beta} = \frac{s - 1}{2\sqrt{s}}$$

where the parameter s is the criterion for the maximum VSWR and ω_+, ω_-, ω_0 represent the higher frequency bound, lower frequency bound and central frequency of the antenna, respectively.

http://dx.doi.org/10.5772/58815

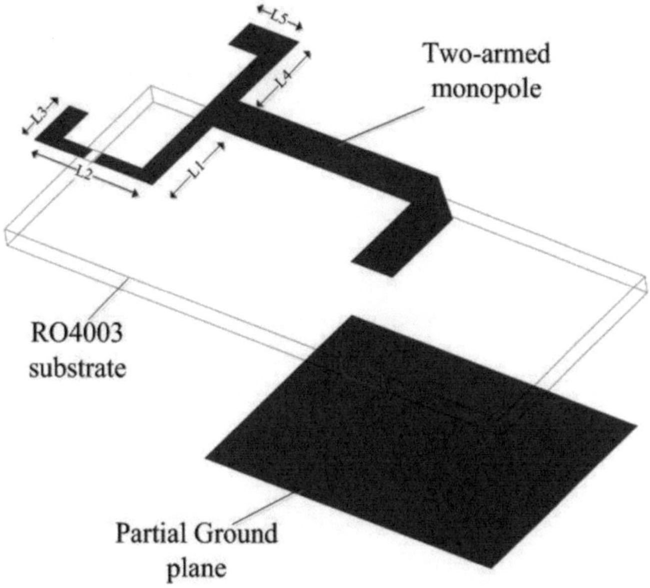

Figure 8. The configuration of the electrically small printed monopole antenna proposed in [7]

The lower bound of the quality factor (Q_{lb}) of the antenna was calculated by using the formula [8]:

$$Q_{lb} = \left[\frac{1}{(ka)^3} + \frac{1}{ka} \right]$$

From Table 1, it can be seen that the antenna has *ka* smaller than 0.5 and has a quality factor very close to its theoretical lower bound, where *a* is the radius of sphere that can enclose the maximum dimension of the antenna and *k* is the wavenumber.

Properties	Proposed Antenna
Frequency (GHz)	2.45
Size (Wmm xLmm)	10×10.5
a (mm)	7.25
ka	0.37
Simulated Radiation efficiency (%)	72
Qlb	15.9
Measured 3:1 VSWR bandwidth (%)	5.10
Calculated Antenna Q	22.6

Table 1. Summary of performance of the proposed printed monopole antenna in [7]

3. Low cost printed planar monopole for mobile terminals

The rapid growth in mobile communications increases the needs of designing multiband internal antennas for mobile terminals. Meanwhile, it is also desirable to design such antennas as compact as possible. Planar Invert-F Antenna (PIFA) is one type of conventional antennas that has been widely employed in mobile phones. In [9, 10], two coupled-fed compact multiband PIFAs for wireless wide area networks (WWAN) were proposed for internal mobile phone antenna applications. The size reduction of these two antennas was achieved by shorting the antenna to the ground and bending the antenna structure. Printed monopole slot antennas and printed loop antennas have also been widely studied for multiband internal mobile phones. In [11, 12], two folded monopole slot antennas that can cover the penta-band WWAN operation were proposed for clam-shell mobile phones. These two antennas were designed by making several slots on the top of the ground plane. In [13-15], printed half-wavelength and meandered loop techniques were proposed for the design of multiband antennas for mobile handsets. However, all of these antennas have operating bands only covering GSM850/900 and DCS/PCS/UMTS bands, which are not enough for nowadays wireless communications. To make the antenna resonant at additional bands including Wireless LAN, one novel PIFA structure combining shorted parasitic patches, capacitive loads and slots was designed to support both quad-band mobile communication and dual-band wireless local area network (WLAN) operations [16]. Although this antenna can operate at several bands, it is extremely difficult to fabricate due to its complex structure. In [17], multiband operation including the WWAN and WLAN 2.4 GHz was achieved by cutting slots of different lengths at the edge of the system ground plane of the mobile phone. Operation in additional bands including GSM/DCS/PCS/UMTS/WLAN/WiMAX were achieved by cutting the loop-like slot on the top of the ground plane and shorting it to the ground plane [18]. However, shorting the antenna to the ground makes the resonant frequencies of the antenna vulnerable to the length of the ground plane. In fact the ground plane size used in [18] is smaller than the size of the system ground plane for a mobile phone. Other techniques have also been developed to design compact multiband antennas for wireless communications. In [19], a multiband antenna that can support WWAN and 2.4 GHz WLAN frequency bands was implemented by using a switchable feed and ground. In [20], a small size multiband antenna for wireless mobile system was designed based on double negative (DNG) zeroth order resonator (ZOR). However, it is noticed that these antennas have rather complex structures and they are quite difficult to fabricate. In [21], a chip inductor was embedded in the printed monopole antenna, which resulted in a compact antenna for mobile handset application.

One compact multiband printed monopole for mobile application has been recently presented in [3]. This design overcomes some of these limitations discussed above. Figure 9 shows the structure of the proposed antenna and the main dimensions of the antenna elements are given in Figure 10. The antenna element is printed on the top side of the substrate while the ground plane is located at the bottom side. Behind the monopole antenna, there is no ground. The chip inductor, of series Coilcraft 0402HP with an inductance of 20 nH, is embedded between the branch A and B as shown in Figure 10. This antenna has a multi-branch structure, each of which determines different resonant frequencies. The lowest resonant frequency, 960 MHz, is determined by both the inductance of the chip inductor and the overall length of branch A and B. Although the chip inductor can also influence the resonant frequency at 1800 MHz to some extent, this resonance is mainly determined by the length of branch A. The overall length of branch D and the length of branch E determine the resonant frequencies at 2.4 and 5.2 GHz, respectively. The frequency band at 3.8 GHz is related to the length of branch C and the width of branch A.

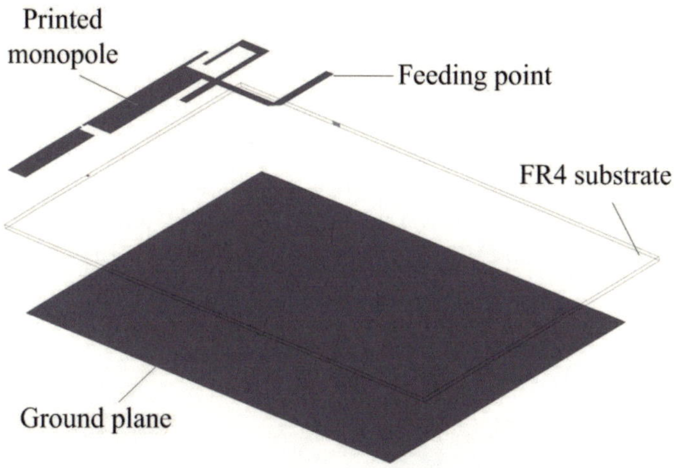

Figure 9. Structure of the multiband printed monopole antenna for mobile terminals [3]

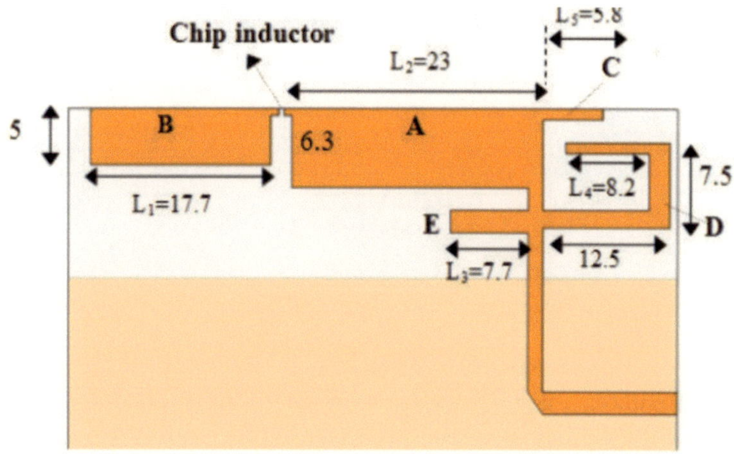

Figure 10. The main dimensions of the multiband printed monopole antenna [3]

This antenna is printed on the inexpensive substrate FR4 (relative permittivity of 4.4) with thickness of 0.8 mm and size 100mm ×60mm, which is a reasonable circuit board size for a PDA or smart phone device. To achieve a better impedance matching at each band, the length and width of each branch of the antenna were optimized through numerical simulations. In the simulation set-up, the model of the chip inductor is built based on the studies presented in [7]. As stated before, the chip inductor mainly influences the first two lower frequency bands. In these two lower frequency bands, the value of the chip inductor is more critical in determining the lowest resonant frequency; as a result, in the simulation set-up, the equivalent inductance and series resistance of the chip inductor model were calculated at 960 MHz using the formulas provided in [7], and they are 20.6 nH and 2Ω, respectively.

Figure 11 shows the measured and simulated reflection coefficient of the proposed antenna. This antenna was measured using the network analyzer Agilent PNA E8363B. It can be observed that there is a good agreement between the measurement and simulation results. The experiment result shows that the proposed antenna has 3:1 VSWR bandwidth covering 860-1060 MHz, 1710-2067 MHz, 2360-2500 MHz, 3250-4625 MHz, 5080-5410 MHz, which includes almost all the required frequency bands for GSM900 (890-960MHz), DCS (1710-1880MHz), PCS (1850-1990MHz), UMTS (1920-2170MHz), WLAN dual band (2400-2484/5150-5350MHz) and WiMAX (3400-3600MHz) operations.

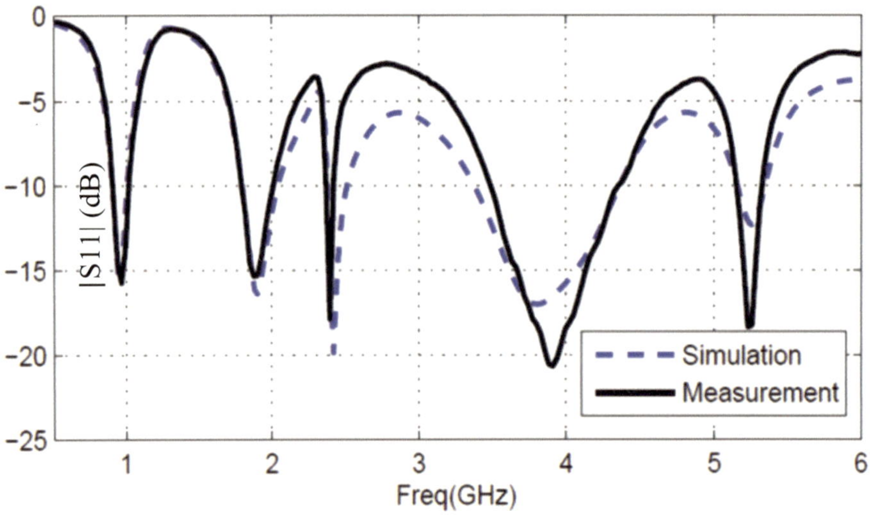

Figure 11. The measured and simulated reflection coefficient of the multiband printed monopole antenna [3]

Figure 12 presents the comparison of the simulated reflection coefficient between the proposed antenna with and without the embedded chip inductor. It is found that without the chip inductor, at the lowest frequency band the antenna can only resonate at around 1.1 GHz. After introducing the chip inductor, this resonant frequency reduces to 960 MHz and also brings down other higher modes to become resonant at 1.8 GHz. It is also observed that the chip inductor has little influence on the resonant frequencies at 2.4 and 5.2 GHz.

Figure 13 shows the simulated surface current distribution of the proposed antenna at each operation frequency. It is observed that at 960 MHz, there is a strong current on branches A

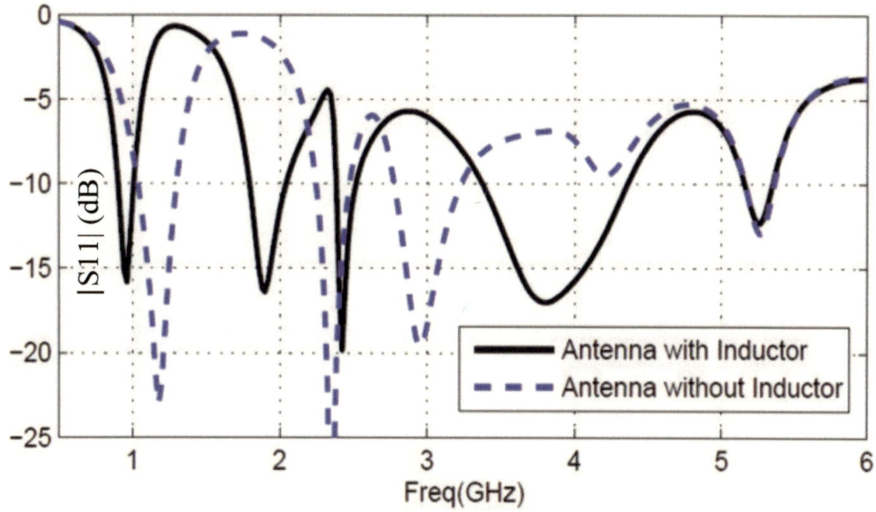

Figure 12. The comparison of the simulated reflection coefficient between the multiband printed monopole antenna with and without the embedded the chip inductor [3]

and B. At 1800 and 1900 MHz, the current is mainly distributed on branch B. It is also clear that branches D and E are responsible for the resonant frequency at 2.4 and 5.2 GHz, respectively. Regarding the resonance at 3.8 GHz, it is mainly determined by branch C and the coupling between branch C and D.

Besides being a completely planar structure, another advantage of the proposed antenna is that the size of the ground plane has little influence on its resonant characteristics compared to the designs that short the antenna structure to the ground plane. The proposed monopole antenna with different lengths of the ground plane has also been investigated. Figure 14 shows the simulated reflection coefficient of the proposed antenna with ground planes of different lengths. It was found that when decreasing the length of the ground plane, at the desired frequency bands the proposed antenna only exhibits small frequency shifts and some changes on the amplitude of the reflection coefficient.

The scenario in which the antenna is put into the center of a plastic housing box was also investigated in this work. In the simulation model, the wall of the plastic housing is 1 mm thick, 14 mm high and has a dielectric permittivity of 3.5. The simulation results (Figure 15) indicate that, compared to the case when the antenna is radiating in free space, there is almost no influence on the reflection coefficient of the proposed antenna except for a small frequency shift at the 3.8 GHz band, when within the plastic housing.

The measured radiation patterns of the proposed antenna in free space are presented in Figure 16. It is found that at all the desired frequencies the proposed antenna has radiation patterns similar to a typical monopole antenna, which normally has omnidirectional radiation patterns. The simulation results also suggest that the antenna has moderate gain and efficiency at its operation frequency bands. Table 2 summarizes the peak gain and radiation efficiency at the desired frequencies. It is observed that at 5.2 GHz, the radiation efficiency is rather low compared to other resonant frequencies. This can be explained by the fact that there is a strong coupling between the branch **B** and **E** (see Figure 12) while the branch **E** is also very closed to the ground plane, with a negative impact on the radiation efficiency.

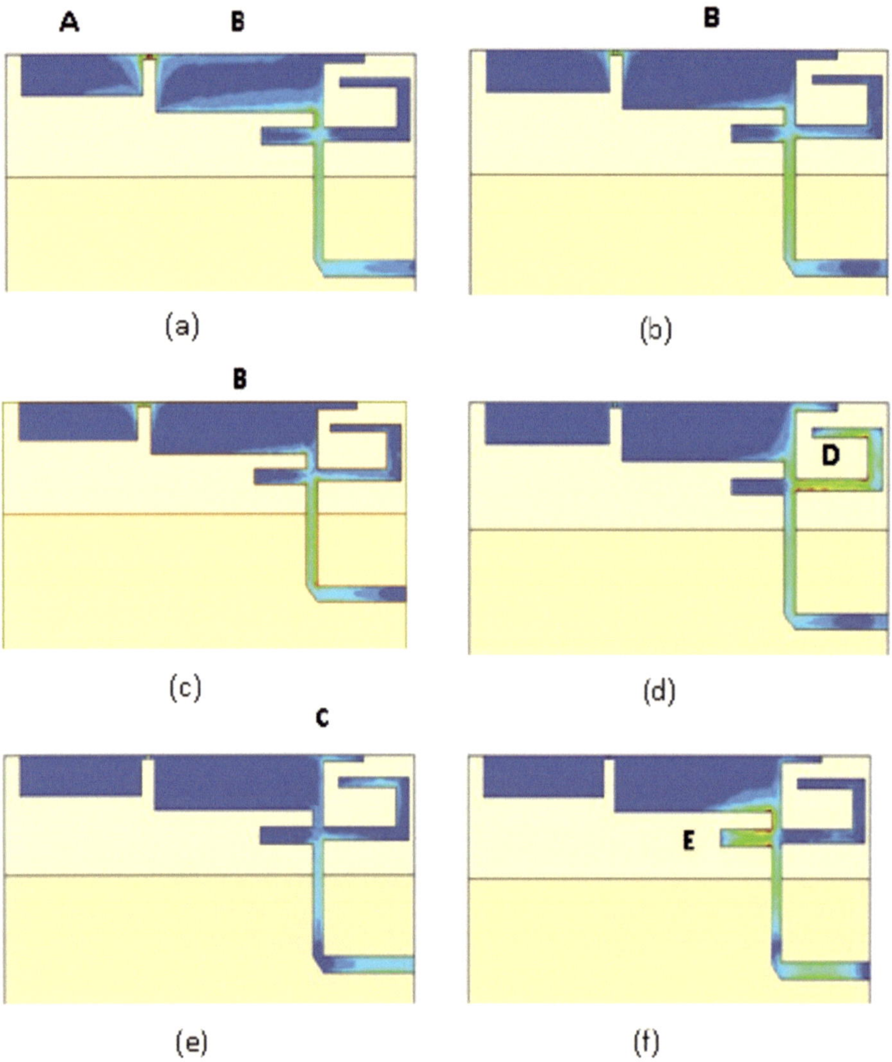

Figure 13. The simulated surface current distribution of the multiband printed monopole antenna [3] : (a) 960 MHz; (b)1800 MHz; (c)1900 MHz; (d)2.4 GHz; (e)3.8 GHz and (f)5.25 GHz. The stronger current is represented by lighter colors

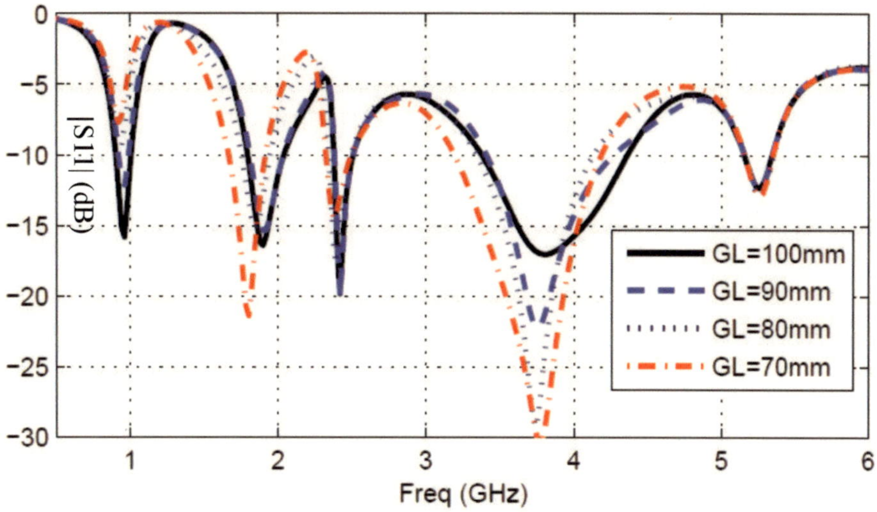

Figure 14. The simulated return loss of the multiband printed monopole antenna with ground planes of different lengths [3]

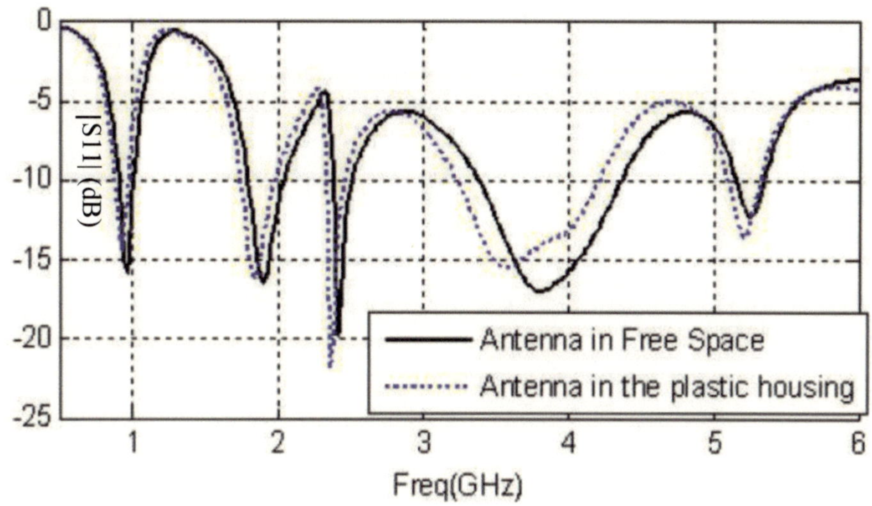

Figure 15. Comparison of the simulated S11 of the multiband antenna when placed in a plastic housing and in free space [3]

The Specific Absorption Ratio is also analyzed in this study. The simulation result indicates that the SAR value averaged over 1 gram of head tissue is 1.4 W/Kg, which meets the released SAR limitation of 1.6 W/Kg. It is expected that the SAR value in reality will be smaller than the simulated one due to the adding of the case for the mobile phones.

Frequency(GHz)	Simulated Peak Gain(dBi) / Radiation Efficiency
0.96	1.5 (93.7%)
1.8	2.6 (91.9%)
1.9	2.5 (92.2%)
2.4	2.7 (77.3%)
3.5	2.8 (86.8%)
5.2	1.7 (67.9%)

Table 2. Simulated Peak gain and radiation efficiency of the proposed antenna at each frequency band [3]

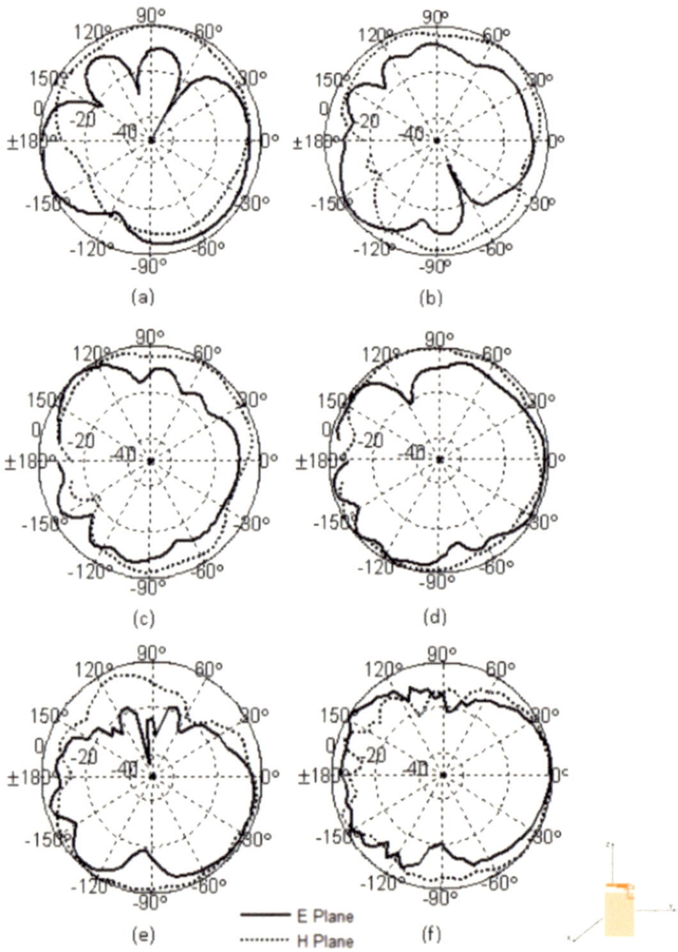

Figure 16. Measured E-plane (X-Z Plane) and H-plane (X-Y plane) radiation patterns of the proposed multiband antenna at: (a) 960MHz; (b) 1800MHz; (c) 1900MHz; (d) 2.4GHz; (e) 3.5GHz; (f) 5.2GHz [3]

4. Small size printed monopole array

4.1. Fractal monopole antenna array

To increase the directivity of an antenna system, the use of antenna arrays is an effective solution if an additional antenna element can be added in the wireless device. A single feed antenna array has the advantages of easy fabrication and does not need any extra RF components such has a phase shifter. It is desirable to design such antenna in a planar structure as it can simplify the fabrication process and reduce the fabrication cost. One compact single feed multiband printed monopole antenna array using the 2nd iteration of the Minkowski fractal geometry designed for WLAN dual band application is presented in [22].

The 2nd iteration of the Minkowski fractal geometry (see Figure 1) was chosen for this design due to its compact size. Figure 17 shows the geometry of the proposed fractal monopole array. This antenna was fabricated on a Roger 4003 substrate of thickness 0.813 mm and relative permittivity 3.38. The substrate is 112 mm long and 65 mm wide, which is the size for a typical PDA terminal. The antenna is constituted by two equal 2nd iteration Minkowski fractal monopoles fed by a single microstrip line of 1.89 mm wide. The line width of the fractal geometries is 0.25 mm and they are connected to the feed line by another horizontal microstrip line of width 1.2 mm. The partial ground plane is printed on the back side of the substrate and the antenna is printed on the top side.

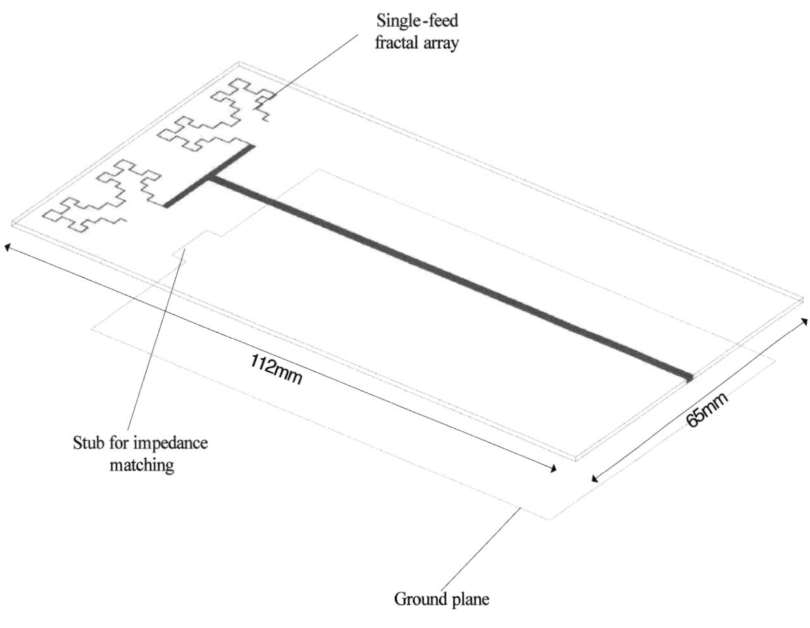

Figure 17. The single-feed fractal monopole array on a PDA size substrate proposed in [22]

As can be seen from Figure 17, on the top size of the partial ground plane, a rectangular stub is added. Without introducing the rectangular stub on the partial ground plane, it is found that the bandwidth of this antenna is not as good as expected: the bandwidth at the higher band (5 GHz) is quite narrow. Therefore, it is necessary to find a method to improve the bandwidth of the antenna at the higher band without affecting too much the resonant frequency at the lower band. Some common impedance matching methods such as quarter-wavelength transformer line or microstrip taper line, besides their large size, they are not suitable for this application, since they can only be applied to single band antennas. After several attempts, it was found that by adding a stub on the top edge of the ground plane, the impedance match of the antenna can be improved with little influence on the original resonant frequencies.

The improvement of the impedance matching of the proposed fractal antenna with the addition of a stub on the partial ground plane can be explained by modeling the stub as an equivalent L-Matching Network, as shown in Figure 18. The value of the inductance (L_1) and capacitance (C_1) at each resonant frequency is determined by the size/shape of the stub and the thickness/permittivity of the substrate. For the antenna presented in [22], due to the use of the fractal geometry, which has the advantage of self-affinity and exhibiting similar radiation characteristics at multiple resonant frequencies, the impedance matching was improved simultaneously at both resonant frequencies with the addition of an equivalent L-Network. This is one additional merit of employing fractals in monopole antenna design.

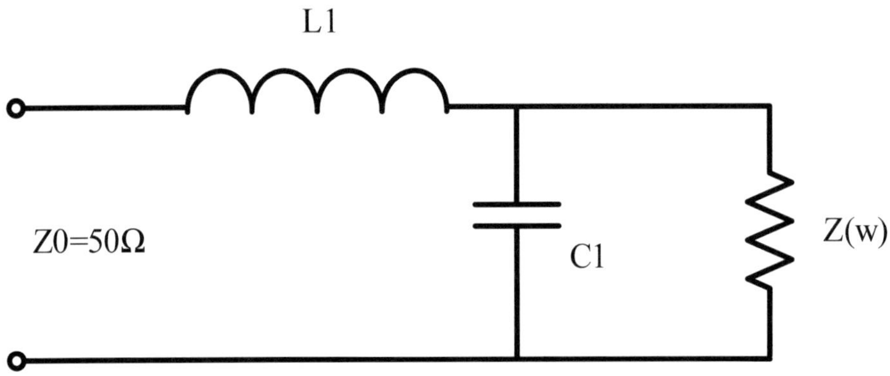

Figure 18. Antenna with an L-Matching Network

Figure 19 shows the measured and simulated reflection coefficient of this design. It can be observed that further optimization is required, to further increase the operating bandwidth of the antenna: it has a 10 dB bandwidth from 2.32 to 2.49 GHz and from 5.1 to 5.88 GHz, which covers the required 2.4, 5.2 and 5.8 GHz bands for 802.11a/b/g applications. Comparing the measured and simulated results, some frequency shifts were observed, which might be caused by the fabrication accuracy or the uncertainty of the dielectric constant of the substrate. By adjusting the size of the fractal geometry, the resonant frequencies can be easily tuned to the desired ones.

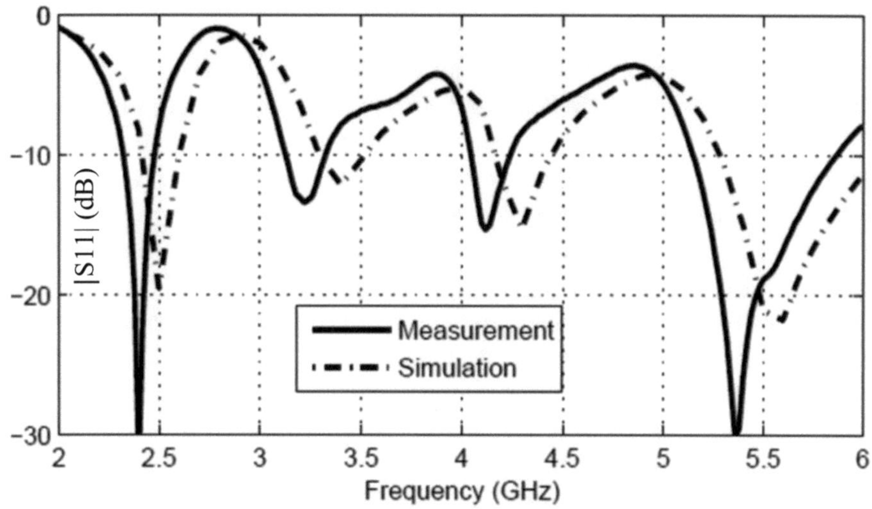

Figure 19. The measured and simulated S11 of the antenna array on a PDA size substrate [22]

The measured radiation patterns for this single-feed array show that at the lower band, the radiation pattern of this antenna array is similar to a normal printed monopole antenna, which has a isotropic radiation pattern at the H plane and two broadside radiation pattern at the E plane. In the upper band, the radiation patterns at both 5.2 and 5.8 GHz are more or less omnidirectional but there are some nulls in the E plane, which are due to the cancellation from the two radiation elements. The measurement results also indicate that the maximum gain of this printed monopole array can reach 2.3 dBi in the lower band and 5.6 dBi in the upper band. Compared to the case of a single radiation element, a minimum of 2 dB gain improvement has been achieved. Based on the simulation results, the radiation efficiency of this antenna array is 86% at 2.4 GHz, 82% at 5.2 GHz and 89% at 5.8 GHz.

4.2. Inverted-L antenna array for MIMO applications

Multiple-Input-Multiple-Output (MIMO) techniques enable a wireless device to transmit or receive data with higher data rate. The recently announced IEEE 802.11n and Long Term Evolution (LTE) standard requires the wireless LAN devices and mobile devices to support MIMO. The use of antenna arrays can improve the diversity performance of the antenna, which in turn increases the channel capacity by reducing the fading, suppressing both the random frequency modulation and co-channel interference. The biggest challenge in designing compact antenna arrays is how to maintain a good isolation between antennas that are closely spaced. To have good space diversity, traditionally the space between each antenna elements is required to be approximately half of the wavelength. However, for most of the commercial wireless devices, it is impossible to follow this rule due to the size constraints. The objective of this work is to explore solutions to design compact antenna arrays. The methodology adopted in this study employs the 'neutralizing technique'.

Designing a WLAN antenna for an USB dongle requires techniques for antenna miniaturization as the available volume left for the antenna is quite small compared to the wavelength at the required resonant frequency, which is quite challenging. As an example, in [23] a USB memory size antenna for 2.4 GHz Wireless LAN (WLAN) was achieved by using a folded trapezoidal antenna. In an USB dongle, the available volume for mounting the antennas is typically around 10×17×5 mm^3. With respect to the design of antenna arrays for USB dongles, it is a challenge task to improve the isolation between each antenna element, since the antennas have to be placed in close proximity. In [24], a dual band two antennas array was proposed. This antenna consists of an L-shape patch and a via trace connecting the via to the ground. To reach the expected performance, it needs precise fabrication and the experimental result shows that the isolation of this antenna array at 2.4 GHz is less than 9 dB. In [25], a MIMO antenna array for mobile WiMAX (3.5 GHz) was presented. This antenna has a 3D structure and the high isolation was achieved by using a common T-shaped ground plane. The disadvantages of this antenna array are that it is difficult to fabricate and the size of the ground plane can have a great effect on the radiation performance of the antenna due to the shorting structure. Regarding the design of compact planar antenna arrays for WLAN 5.8 GHz on a USB dongle, research has shown that there are few publications in this area, which is the main motivation behind this work. Recently, a new method named Neutralization Techniques has been proposed [26]. Using this method, the isolation of two Planar Inverted-F Antennas (PIFAs) can be improved through neutralizing the current of two antennas without the need of adding extra space for antenna design. So far, this method has only been applied in the design of PIFA antennas and there are few studies investigating the use of the neutralization technique. In this work, we further investigate this technique in the design of an Inverted-L antenna (ILA) array.

In [27], a compact and low cost Inverted-L antenna array is proposed for the MIMO application. Figure 20 shows the structure of a classic ILA. The ILA can be viewed as a bent monopole antenna and the total length of the inverted-L, L1+L2, needs to be approximately one quarter of wavelength at the resonant frequency of interest. However, the challenge of this work is that the two antennas need to be closely located in a small area of an USB dongle.

Figure 21 presents the structure of the proposed ILAs. The proposed antenna is fabricated on 0.8 mm thick FR4 with relative permittivity of 4.4 and loss tangent of 0.02. The distance between the two feeding points is $0.15\lambda_{5.8GHz}$ and the gap (d_1) between these two antennas is only $0.02\lambda_{5.8GHz}$, where $\lambda_{5.8GHz}$ represents the free space wavelength at 5.8 GHz. This antenna array has two equal ILAs that are located within a small distance on the PCB board of the USB dongle. Based on the concept proposed in [26], a neutralizing line is added between the two antenna elements to increase the isolation. The length of the neutralizing line is critical in determining the frequency band where the isolation between the two antenna ports can be improved. Increasing the length of the neutralizing line can make the antenna array to have good isolation at the lower frequency band. According to [26], the location of the neutralizing line needs to be placed where the surface current is maximum (minimum E field) and the length of it needs to be approximately a quarter wavelength.

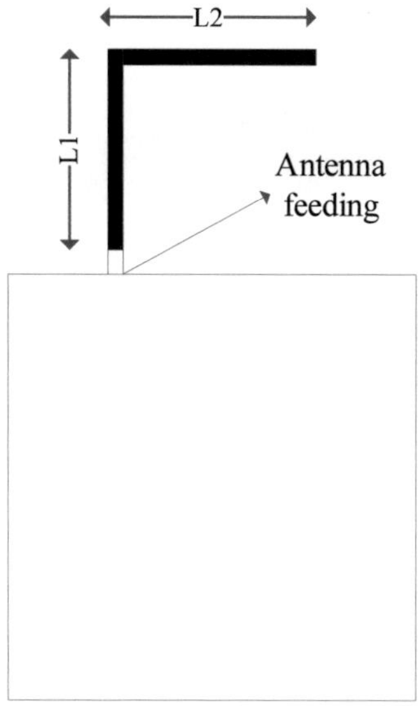

Figure 20. The structure of a typical Inverted-L antenna [27]

It is found in [27] that this antenna structure exhibits poor impedance matching at the desired frequency. The low input impedance of the ILA antenna is in fact one of its disadvantages [28]. The typical method employed to solve this problem for an ILA is to short the antenna element to the ground plane and change the feeding position, which in turn increases the input impedance of the antenna. Then the antenna becomes an Inverted-F antenna (IFA), whose input impedance is easier to be matched. However, shorting the antenna to the ground plane will increase the impact of the ground plane size to the radiation performance of the antenna. When connecting the USB dongle to a PC, for example, the equivalent size of the ground plane for the antenna is extended. In this scenario, the antenna may fail to operate at the desired frequency band. Moreover, the isolation between the antennas may also be influenced by shorting them to a common ground plane. Therefore, it is better to solve this limitation without resorting to short the antenna to the ground. Instead, the technique proposed in [27] improves the impedance matching of the antenna array by including one vertical stub in the middle of the neutralizing line, as depicted in Figure 21. From the aspect of the antenna array, where the isolation between the antennas is of concern, adding this stub has little influence on the isolation between the two antennas as the isolation is mainly controlled by the length, width and position of the horizontal neutralizing line. Meanwhile, for the single antenna itself, the equivalent antenna structure is one bent monopole with an L-shape stub, which operates as an impedance transformer.

Figure 22 shows the measured reflection coefficient and isolation of the proposed antenna array. The measurement results suggest that the proposed ILA array has a 10 dB return loss bandwidth from 5.7 to more than 6 GHz, which is more than the specification required for the WLAN 5.8 GHz frequency band of interest (5.725 to 5.875 GHz). This makes the performance of the proposed antenna more robust during product integration, such as immunity to

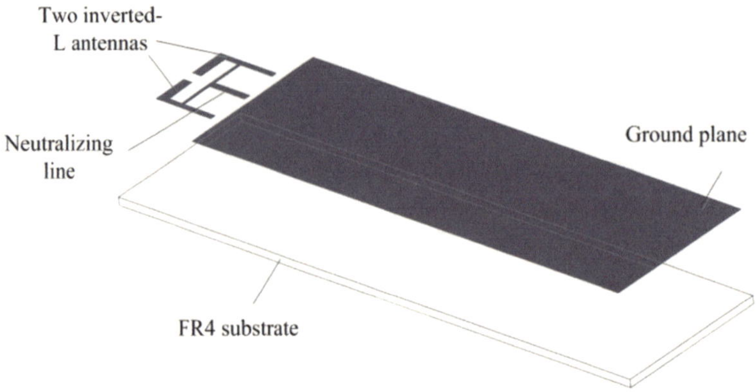

Figure 21. The structure of the proposed ILAs array in [27]

proximity to other components and the product enclosure, thus providing some margin against proximity effects which can lead to some frequency shifts. It is also found that the isolation between the two antennas is always better than 10 dB from 5.5 to 6.0 GHz and within the desired WLAN operation band, an isolation of 12 dB or more is obtained. It is observed that there is some frequency differences (less than 100 MHz) between the measured reflection coefficients of the two ports of the antenna array. This is due to the fabrication accuracy and soldering of the feeding cable, which results in the asymmetrical response of the two antenna elements. The measured results indicate that the proposed antenna array exhibits a maximum gain around 2.5 dBi at 5.8 GHz.

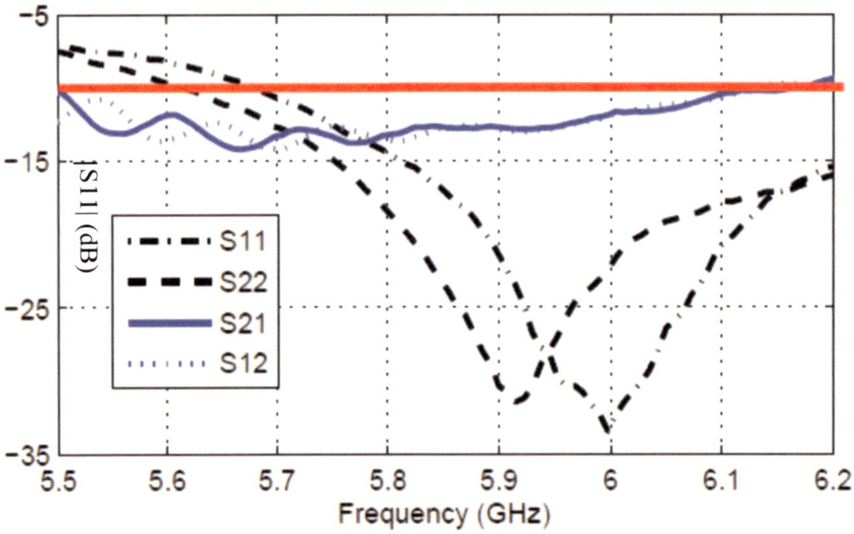

Figure 22. Measured reflection coefficient of the ILAs array proposed in [27]

5. Recent development

It has been shown in this chapter that the usage of two strips can contribute to the design of a dual band printed monopole. Recent research in [29] demonstrates that instead of using only two strips, a triple-band monopole can be designed using three strips. The structure of the triple-band printed monopole in presented in Figure 23. As seen, the printed monopole has three strips, each of which corresponds to a resonant frequency. This implies that introducing multiple strips, a multiband printed monopole can be obtained. This is one of the advantages of the printed monopole compared to other types of antennas. However, the difficulty of this approach lies in how to match the monopole at different resonant frequencies and reduce the influence from the mutual couplings between different strips.

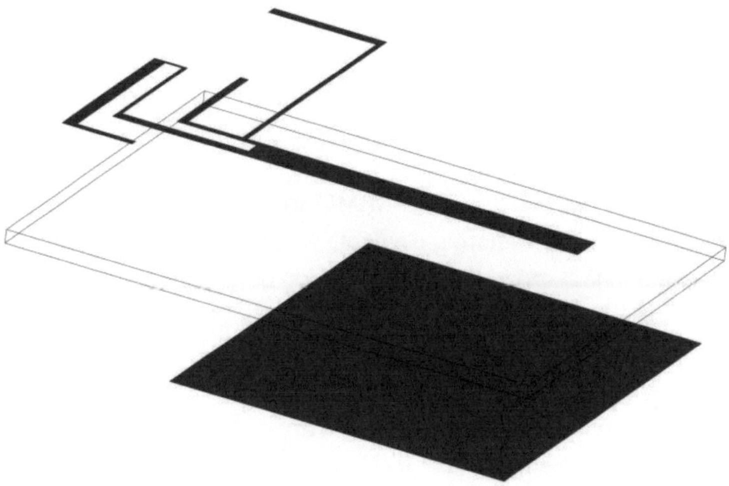

Figure 23. The proposed triple band printed monopole by [29]

It is also known that either using the meandered lines or the introduction a chip inductor can contribute to the size reduction of the monopole. Recently, a broadband LTE/WWAN antenna was designed for the tablet PC application [30]. This monopole is designed by employing both the meandered lines and chip inductor, thus greatly reducing the size of the antenna, whilst reaching a multiple frequency band operation. Similarly, a small printed monopole for DVB application is achieved by introducing a varactor on the meander line monopole [31]. By adding the varactor on the antenna radiating element, not only the antenna size is reduced, but also the frequency reconfigurablility can be achieved. Therefore, it can be concluded that combining different antenna miniaturization techniques to design a small printed monopole is an effective approach. Figure 24 demonstrates one example of introducing a lumped element on a meander line structure.

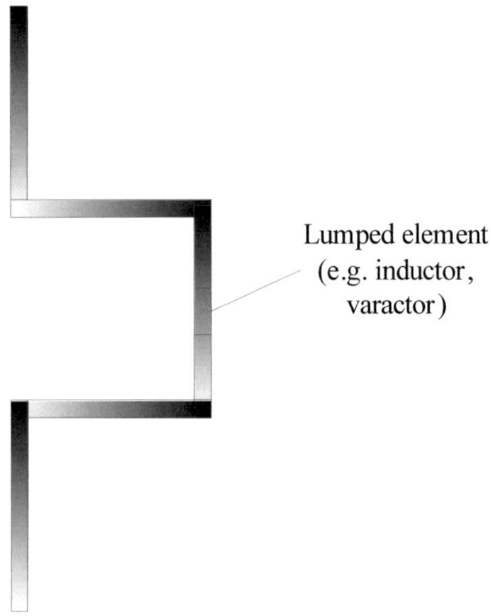

Figure 24. Demonstration of adding a lumped element on the meander line

To design a compact monopole array for the MIMO application, the key issue is to keep a high isolation between two or more radiating elements when they are closely spaced. The simulation results presented in [32] show that the orientation of the antenna elements can be critical in determining the isolation between the antennas. For example, with the spacing between two antennas elements of only one tenth of the wavelength, when the two monopoles are orthogonally oriented as the one shown in Figure 25, about 10-dB improvement in isolation can be observed. This can be explained by the polarization diversity. However, this approach will not be practical if a single polarization for the receiving signal is required to be used.

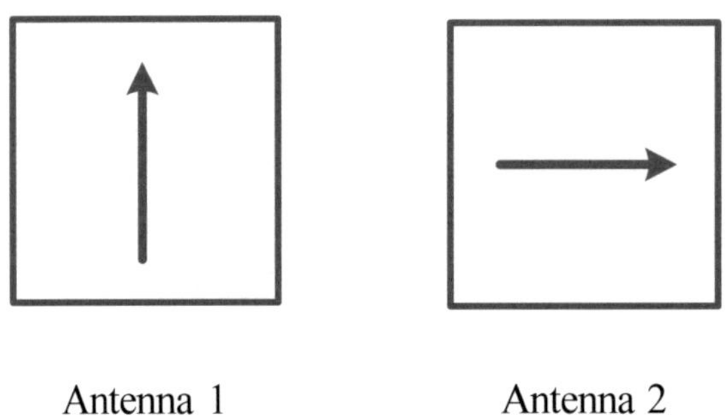

Figure 25. Example of placing two antennas in orthogonal position

Recently, a new approach based on the neutralizing technique, which has been introduced in this chapter, is proposed in [33]. Figure 26 shows the configuration of the neutralizing line. In this approach,a high isolation over a wideband frequency is reached by creating four current paths between the two antenna elements, which is achieved by attaching the neutralizing line to both the antenna elements and the feed line at its maximum current position. The measurement results provided in [33] show that this is an effective method to design a compact wideband printed monopole array for MIMO applications. This constitutes a further development for the conventional neutralizing technique proposed by [26].

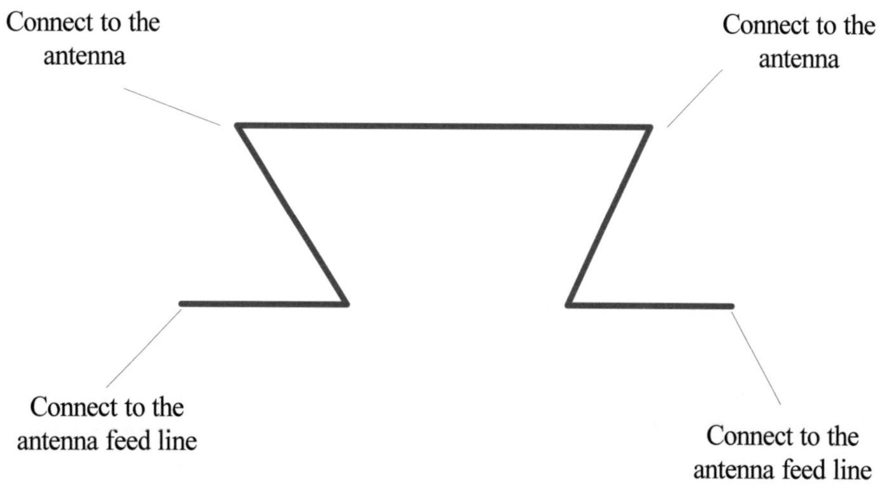

Figure 26. The modified the neutralizing line proposed in [31]

6. Conclusion

In this chapter, several techniques that can be employed to design compact and low cost printed monopole antennas and antenna arrays have been introduced. Either using some special geometry (e.g. fractals) or introducing lumped elements on the radiating elements, multiband monopole with reduced size can be implemented. The recent studies also show that combining both methods, additional size reduction can be achieved. However, the disadvantage is that the antenna radiation performance will be influenced. In the field of compact printed monopole arrays, the use of the neutralizing technique has been proved to be an effective method that can be applied to the planar monopole design, which can result in a simple and low cost solution. To reach a wideband operation, the modified neutralizing line proposed in [31] provides a good solution.

Author details

Qi Luo[1], Jose Rocha Pereira[2,3] and Henrique Salgado[4,5]

1 School of Engineering and Digital Arts, University of Kent, Canterbury, UK

2 University of Aveiro, Portugal

3 Instituto de Telecomunicações, Aveiro, Portugal

4 Faculdade de Engenharia Universidade do Porto, Porto, Portugal

5 INESC TEC - Instituto de Engenharia de Sistemas e Computadores do Porto, Porto, Portugal

References

[1] L. J. Chu, "Physical Limitations of Omni-Directional Antennas," *Journal of Applied Physics,* vol. 19, no. 12, pp. 1163-1175, 1948.

[2] B. B. Mandelbrot, *The fractal geometry of nature,* Updated and augm. ed., New York: W.H. Freeman, 1983.

[3] Q.Luo, "Design synthesis and miniaturization of multiband and reconfigurable microstrip antenna for future wireless applications," University of Porto, 2014.

[4] Q. Luo, H. M. Salgado, and J. R. Pereira, "Fractal Monopole Antenna Design Using Minkowski Island Geometry," *2009 Ieee Antennas and Propagation Society International Symposium and Usnc/Ursi National Radio Science Meeting, Vols 1-6,* pp. 2639-2642, 2009.

[5] Q. Luo, J. R. Pereira, and H. M. Salgado, *Inverted-L Antenna (ILA) Design Using Fractal for WLAN USB Dongle,* 2013.

[6] N. Cohen, "Fractal antenna applications in wireless telecommunications." pp. 43-49.

[7] Q. Luo, J. R. Pereira, and H. M. Salgado, "Compact Printed Monopole Antenna With Chip Inductor for WLAN," *Ieee Antennas and Wireless Propagation Letters,* vol. 10, pp. 880-883, 2011.

[8] A. D. Yaghjian, and S. R. Best, "Impedance, bandwidth, and Q of antennas," *Antennas and Propagation, IEEE Transactions on,* vol. 53, no. 4, pp. 1298-1324, 2005.

[9] K. L. Wong, and C. H. Huang, "Compact multiband PIFA with a coupling feed for internal mobile phone antenna," *Microwave and Optical Technology Letters,* vol. 50, no. 10, pp. 2487-2491, Oct, 2008.

[10] C. H. Chang, K. L. Wong, and J. S. Row, "Coupled-Fed Small-Size Pifa for Penta-Band Folder-Type Mobile Phone Application," *Microwave and Optical Technology Letters,* vol. 51, no. 1, pp. 18-23, Jan, 2009.

[11] C. I. Lin, and K. L. Wong, "Printed monopole slot antenna for penta-band operation in the folder-type mobile phone," *Microwave and Optical Technology Letters,* vol. 50, no. 9, pp. 2237-2242, Sep, 2008.

[12] F. H. Chu, and K. L. Wong, "Simple Folded Monopole Slot Antenna for Penta-Band Clamshell Mobile Phone Application," *Ieee Transactions on Antennas and Propagation,* vol. 57, no. 11, pp. 3680-3684, Nov, 2009.

[13] K. L. Wong, and C. H. Huang, "Printed loop antenna with a perpendicular feed for penta-band mobile phone application," *Ieee Transactions on Antennas and Propagation,* vol. 56, no. 7, pp. 2138-2141, Jul, 2008.

[14] C. I. Lin, and K. L. Wong, "Internal meandered loop antenna for GSM/DCS/PCS multiband operation in a mobile phone with the user's hand," *Microwave and Optical Technology Letters,* vol. 49, no. 4, pp. 759-765, Apr, 2007.

[15] Y. W. Chi, and K. L. Wong, "Half-wavelength loop strip capacitively fed by a printed monopole for penta-band mobile phone antenna," *Microwave and Optical Technology Letters,* vol. 50, no. 10, pp. 2549-2554, Oct, 2008.

[16] P. Ciais, R. Staraj, G. Kossiavas *et al.*, "Compact internal multiband antenna for mobile phone and WLAN standards," *Electronics Letters,* vol. 40, no. 15, pp. 920-921, Jul 22, 2004.

[17] C. I. Lin, and K. L. Wong, "Printed monopole slot antenna for internal multiband mobile phone antenna," *Ieee Transactions on Antennas and Propagation,* vol. 55, no. 12, pp. 3690-3697, Dec, 2007.

[18] H. Hsuan-Wei, L. Yi-Chieh, T. Kwong-Kau *et al.*, "Design of a Multiband Antenna for Mobile Handset Operations," *Antennas and Wireless Propagation Letters, IEEE,* vol. 8, pp. 200-203, 2009.

[19] A. C. K. Mak, C. R. Rowell, R. D. Murch *et al.*, "Reconfigurable Multiband Antenna Designs for Wireless Communication Devices," *Antennas and Propagation, IEEE Transactions on,* vol. 55, no. 7, pp. 1919-1928, 2007.

[20] J. K. Ji, G. H. Kim, and W. M. Seong, "A Compact Multiband Antenna Based on DNG ZOR for Wireless Mobile System," *Ieee Antennas and Wireless Propagation Letters,* vol. 8, pp. 920-923, 2009.

[21] K. L. Wong, and S. C. Chen, "Printed Single-Strip Monopole Using a Chip Inductor for Penta-Band WWAN Operation in the Mobile Phone," *Ieee Transactions on Antennas and Propagation,* vol. 58, no. 3, pp. 1011-1014, Mar, 2010.

[22] L. Qi, H. M. Salgado, and J. R. Pereira, "Printed fractal monopole antenna array for WLAN." pp. 1-4.

[23] K. Sekine, and H. Iwasaki, "USB Memory Size Antenna for 2.4 GHz Wireless LAN and UWB." pp. 1173-1176.

[24] A. Gummalla, M. Achour, G. Poilasne *et al.*, "Compact Dual-Band Planar Metamaterial Antenna Arrays for Wireless Lan," *2008 Ieee Antennas and Propagation Society International Symposium, Vols 1-9*, pp. 4595-4598, 2008.

[25] K. Jaesoon, K. Dongho, L. Youngki *et al.*, "Design of a MIMO antenna for USB dongle application using common grounding." pp. 313-316.

[26] C. Luxey, "Design of multi-antenna systems for UMTS mobile phones." pp. 57-64.

[27] L. Qi, C. Quigley, J. R. Pereira *et al.*, "Inverted-L antennas array in a wireless USB dongle for MIMO application." pp. 1909-1912.

[28] Z. N. Chen, "Note on impedance characteristics of L-shaped wire monopole antenna," *Microwave and Optical Technology Letters,* vol. 26, no. 1, pp. 22-23, 2000.

[29] C. H. Ku, L. K. Li, and W. L. Mao, "Compact Monopole Antenna with Branch Strips for Wlan/Wimax Operation," *Microwave and Optical Technology Letters,* vol. 52, no. 8, pp. 1858-1861, Aug, 2010.

[30] S. H. Chang, and W. J. Liao, "A Broadband LTE/WWAN Antenna Design for Tablet PC," *Ieee Transactions on Antennas and Propagation,* vol. 60, no. 9, pp. 4354-4359, Sep, 2012.

[31] M. Komulainen, M. Berg, H. Jantunen *et al.*, "Compact varactor-tuned meander line monopole antenna for DVB-H signal reception," *Electronics Letters,* vol. 43, no. 24, pp. 1324-1326, 2007.

[32] H. F. Abutarboush, R. Nilavalan, S. W. Cheung *et al.*, "Compact Printed Multiband Antenna With Independent Setting Suitable for Fixed and Reconfigurable Wireless Communication Systems," *Ieee Transactions on Antennas and Propagation,* vol. 60, no. 8, pp. 3867-3874, Aug, 2012.

[33] C. H. See, R. A. Abd-Alhameed, Z. Z. Abidin *et al.*, "Wideband Printed MIMO/Diversity Monopole Antenna for WiFi/WiMAX Applications," *Ieee Transactions on Ant*ennas and Propagation, vol. 60, no. 4, pp. 2028-2035, Apr, 2012.

Chapter 4

Miniature Antenna with Frequency Agility

L. Huitema and T. Monediere

Additional information is available at the end of the chapter

http://dx.doi.org/10.5772/58838

1. Introduction

The need of both mobility and communication leads to the integration of antennas in miniature devices so far non-connected (particularly in medical areas). The dedicated volume for the antenna, including its ground plane, has to be kept at its acceptable minimum, involving low bandwidth. Moreover, due to their poor impedance bandwidth, small antennas tend to be very sensitive to the environment. Indeed, they are directly affected by their immediate surroundings, which disturb their working band, their radiation and their performances [1]. To counter the low bandwidth of the antenna and to adapt it to variable conditions and surroundings, it can integrate active components.

Thus, active components become highly suitable for the development of modern wireless communications. Indeed, they allow the miniaturization, shifting the antenna working frequency to be matched over a wide bandwidth by covering only the user channel and the adaptation of antennas to variable operating conditions and surroundings. It is in this framework that authors will propose in this chapter to detail the integration of active components in antennas to be more compact, smart and integrated.

The first part will address an overview of the most common used techniques for compact antennas to become active. In this goal, active antennas state-of-the-art will be presented:

- The first sub-section will present antennas integrating tunable components such as varactor diodes, MicroElectroMechanical systems (MEMS), Positive Intrinsic Negative (PIN) diode and Field Effect Transistor (FET).
- The second sub-section will focus on active antennas using tunable materials properties, i.e. ferroelectric materials and liquid crystal.

A second part will show relevant parameters for active antennas studies. It will exhibit both challenges and how to integrate active components in order to maximize the antenna performances and efficiency. This part will be supported by concrete examples. Therefore, depending on their intended applications, readers will be prepared to find the best trade-offs between the agility method, the miniaturization and antenna performances.

The last part will be dedicated to present limitations of actual and most common solutions proposed for active and compact antennas. In this framework, new approaches will be detailed to overcome these physical limitations.

2. Overview of compact active antennas

Very small size antennas are needed for future dense wireless network deployment, for example in WBAN (Wireless Body Area Network) where the size is limited to dimensions much smaller (hearing aid, implants) than wavelength (λ_0=12.2 cm at 2.45GHz) or for the DVB-H (Digital Video Broadcasting – Handheld) application, where the miniaturization aspect is even more critical because it is a low frequency standard (λ_0=60 cm at 470 MHz). Therefore, the antenna has to be carefully optimized with trading off fundamental size limitations with its characteristics (especially bandwidth and efficiency). In addition, the environment of miniature antennas will be highly variable resulting in large antenna impedance and propagation channel changes. Several challenges have to be addressed. One of them is the adaptive antenna technology for compensating both low bandwidths and detuning effects.

The most commonly cited performance criterion is the achievable frequency tuning range (TR) defined as:

$TR(\%)=\frac{2(f_{max}-f_{min})}{f_{max}+f_{min}}.100$ where f_{max} and f_{min} are respectively the upper and the lower antenna operating frequency. The frequency tuning can either be continuous or discrete. The continuous frequency tuning is able to continuously cover each channel of a same standard while the discrete frequency tuning can only switch between different standards. This part will present the most common methods to target frequency tunable antennas design.

2.1. Integration of active components

2.1.1. Varactor diodes

For continuous frequency tuning, the integration of varactor diodes within an antenna is the most common approach [2-5]. P. Bhartia et al. were the first to publish antenna integrating varactor diodes [6]. Indeed, they presented a microstrip patch antenna with varactor diodes at the edges of the structure, as illustrated Figure 1. Both rectangular and circular tunable patches were studied, results reveal that 22% and 30% of bandwidth can respectively be achieved by varying the DC-bias-voltage between 0V and 30V.

http://dx.doi.org/10.5772/58838

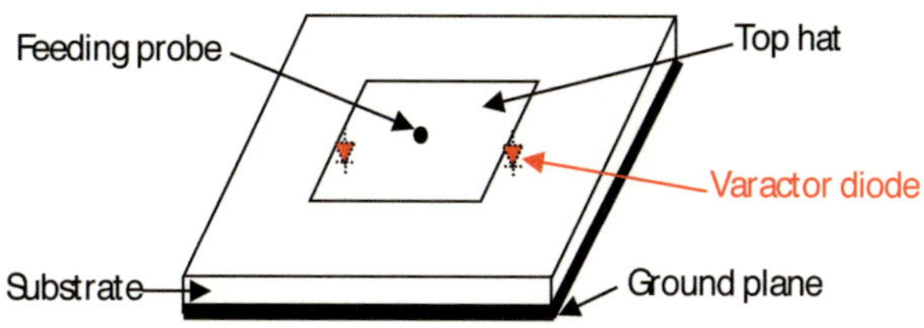

Figure 1. Patch antenna integrating varactor diodes [28]

Slot antennas are also good candidates for the frequency agility [7-9]. N. Behdad and K. Sarabandi presented in [9] a dual band reconfigurable slot antenna. The schematic of its proposed dual-band slot antenna is shown in Figure 2. Matching is performed by choosing appropriately the location of the microstrip feed and the length of the open circuited line.

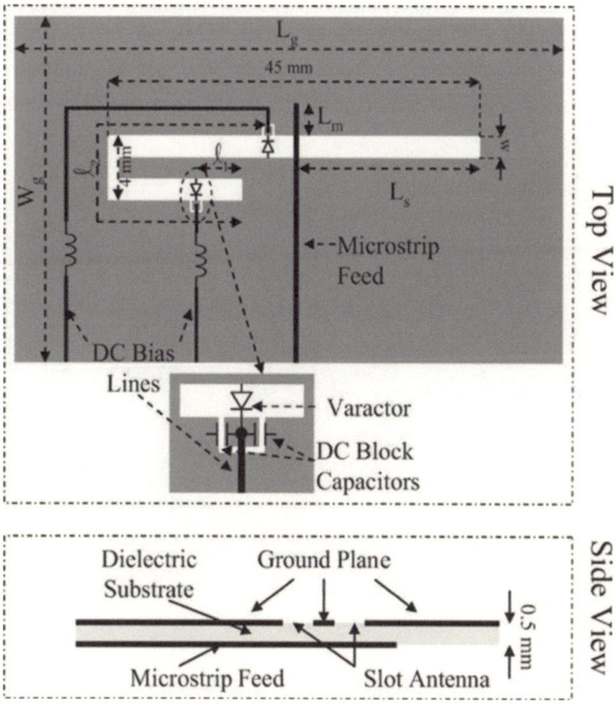

Figure 2. Dual reconfigurable slot antenna [9]

Figure 3 shows the simulated and measured dual-band responses of the antenna where by applying the appropriate combination of bias voltages (V_1 and V_2) the frequency of the first band is kept fixed and that of the second band is tuned. Similarly, as shown in Figure 3, it is possible to keep the frequency of the second band stationary and sweep the frequency of the first band.

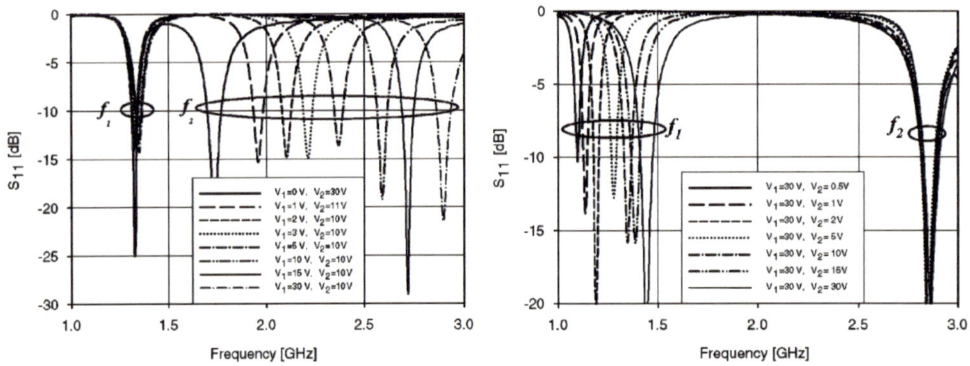

Figure 3. Measured $|S_{11}|$ parameters for different combination of bias voltages [9]

Another example show a 3D Inverted F Antenna [10] designed to cover the entire DVB-H band going from 470 MHz to 862 MHz. To be integrated in a mobile handheld device, the antenna allocated volume had to be very compact. A good trade-off between small sizes and the impedance bandwidth was to choose a structure based on the IFA design. Indeed, the radiating monopole of this kind of structure can be folded all around a material to become more compact (see Figure 4).

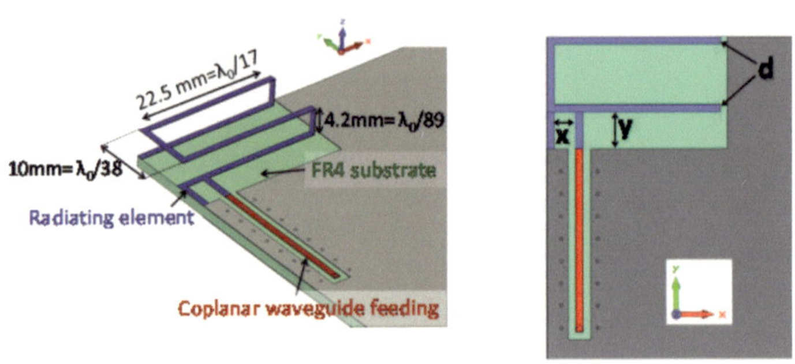

(a) (b)

Figure 4. Inverted F Antenna design (a). Antenna top view (b) [10]

However, the more compact the antenna is, the lowest the bandwidth is becoming. To counter this issue, it has been proved that using a magneto-dielectric material rather than a dielectric one allows enhancing the input impedance bandwidth. Basing on this antenna design, the idea was to integrate a varactor diode to tune the impedance matching all over the DVB-H band (see Figure 5).

A prototype of the tunable antenna has been realized (Figure 6) and measured. For the diode polarization, a DC bias Tee is optimized with SMD components and measured on the DVB-H band (Figure 6). After being validated, it is integrated upstream from the antenna structure as

http://dx.doi.org/10.5772/58838

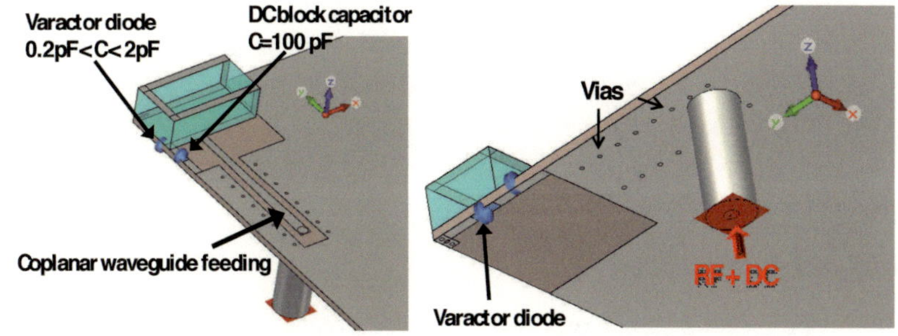

Figure 5. Design integrating both magneto-dielectric material and a varactor diode [10]

shown Figure 6. In order to improve the quality and the reliability of wireless links, the final mobile device is integrating two antennas (see Figure 6) for diversity operations.

Figure 6. DC bias Tee with SMD components and its integration upstream from the antenna (a), integration of two antennas in the tablet dedicated to the DVB-H reception (b) [10]

Figure 7 presents respectively the variation of the input impedances and $|S_{11}|$ parameters of the antenna versus frequency for different values of the varactor diode bias voltages.

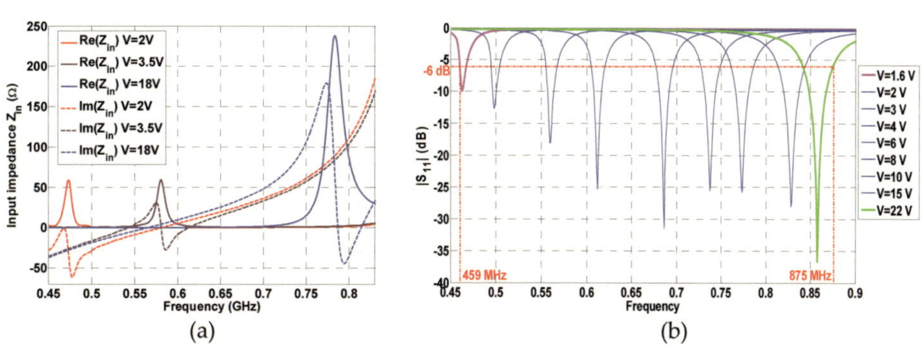

Figure 7. Measured input impedances (a) and $|S_{11}|$ parameters (b) for several DC bias voltages [10]

Therefore, the antenna working band is continuously tuned all over the whole DVB-H band. In the worst case, i.e. for a 2V DC bias voltage, the antenna is matched with $|S_{11}|$ <-6dB in a bandwidth which is covering more than one channel of the DVB-H band at-6 dB (largely suitable for the DVB-H standard).

2.1.2. Positive Intrinsic Negative (PIN) diodes

PIN diodes are using as switches:

- The ON state of the diode can be modelled by a zero resistance, i.e. a continuous metal strip across the slot where the diode is integrated.
- The OFF state of the diode can be modelled as an infinite resistance. The radiating length after the diode is not seen from RF point of view. The effective length of the antenna, and hence its operating frequency, is changing compared with the ON state case.

Selected antenna's types for integrating PIN diodes are often slot antennas or printed antennas (e.g. printed monopole, Inverted F Antenna, …). J-M. Laheurte presented in [11] a slot antenna including pin diodes for multi-frequency operation within a frequency octave.

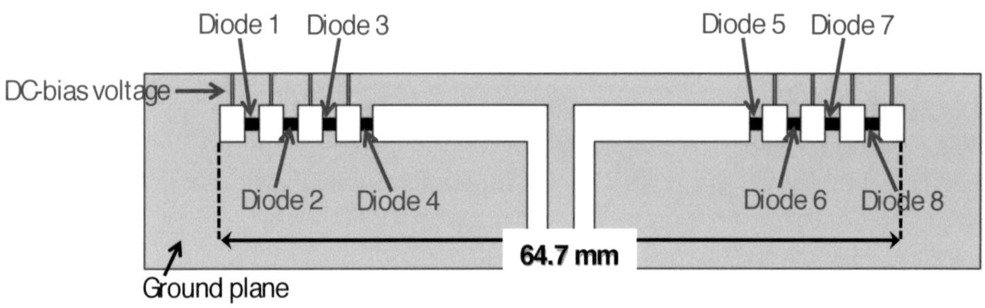

Figure 8. Switchable slot antenna including eight pin diodes [11]

As shown Figure 8, this antenna integrates eight PIN diodes and according to their ON or OFF states combination, the antenna can operate at different and discrete frequency bands (see Figure 9). Instantaneous impedance bandwidths are between 8% and 21% depending on the diodes' states combination.

Eventually, this antenna presents somewhat large dimensions since its main size is higher than à $\lambda_0/2$ at 2.8 GHz. The literature presents smaller antennas integrating PIN diodes since their main size are lower than à $\lambda_0/2$ at the working frequency [12].

Peroulis et al. [13] presented a tunable single-fed S-shaped slot loaded with a series of four PIN diodes. The effective length modification allows this antenna to operate in one of four selectable frequency bands between 530 and 890 MHz.

Before directly studying the tunable slot antenna, both single S-shaped slot antenna and PIN diodes were separately presented and studied. By this way, the issue related to the design of

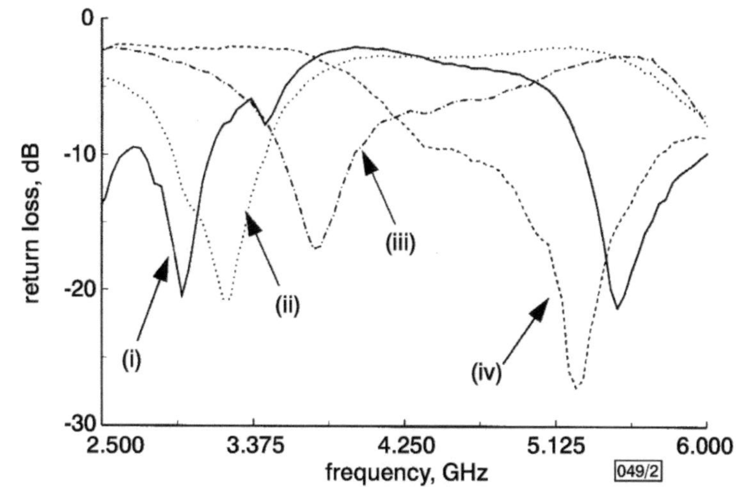

Figure 9. Measured $|S_{11}|$ parameters for different states of diodes: all diodes OFF (i), diodes 1, 8 ON (ii), diodes 1, 2, 7, 8 ON (iii) and diodes 1, 2, 3, 6, 7, 8 ON (iv) [11]

a suitable PIN switch has been grasped before integrate it and show its effects on the antenna performances. Indeed, to implement the electronic reconfigurability, the ideal shunt switches must be replaced by real PIN diodes. Therefore, the RF equivalent circuit of the diode has been studied (see Figure 10) for both the ON and OFF states. The reactive components C_p and L_p are modeling the packaging effect, while the others come from the electric properties of the diode junction in the ON and OFF positions. Then, the switch bias network was presented as an inductor of 470 nH and three 10 pF capacitors (Figure 10).

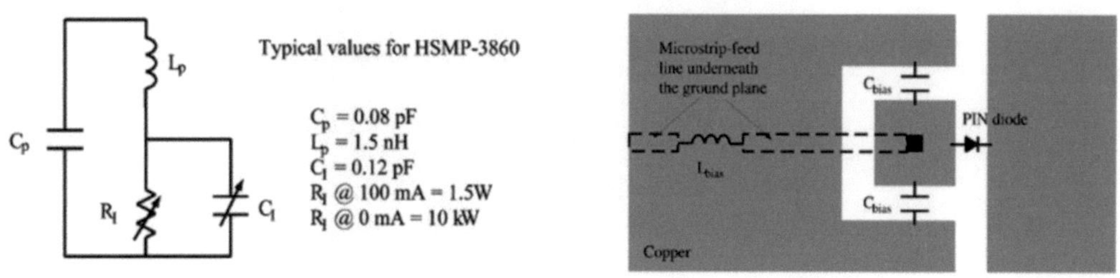

(a) (b)

Figure 10. RF equivalent circuit of the PIN diode (a) and the switch bias network (b) [13]

Finally, a reconfigurable slot antenna design (Figure 11) is presented in this paper. Four switches are used in order to tune the antenna over a range of 540–950 MHz. The integration of the fours PIN diodes allows choosing the operating frequency of the antenna (Figure 11).

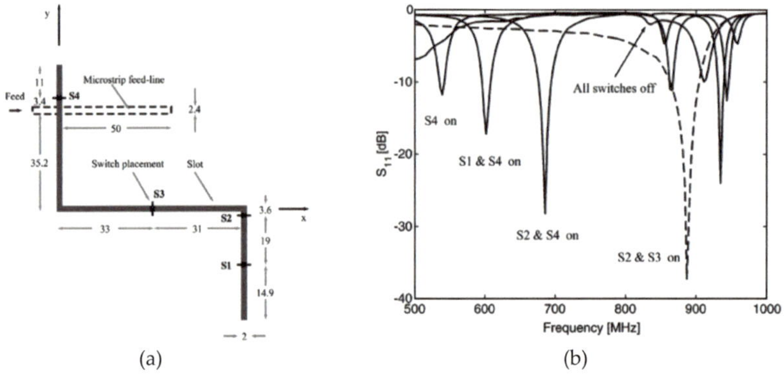

Figure 11. Reconfigurable slot antenna (a) and its measured $|S_{11}|$ parameters (b) [13]

2.1.3. *MicroElectroMechanical systems (MEMS)*

MEMS components can allow:

- Continuous frequency tuning when they are used as a variable capacitance.
- Discrete frequency tuning, when they are used as switches.

E. Erdil presents in [14] a reconfigurable microstrip patch antenna integrating RF MEMS capacitor for continuously tuning the resonant frequency (see Figure 12). The reconfigurability of the operating frequency is obtained by loading one of the radiating edges of the microstrip patch antenna with a CPW stub on which RF MEMS bridge type capacitors are periodically placed.

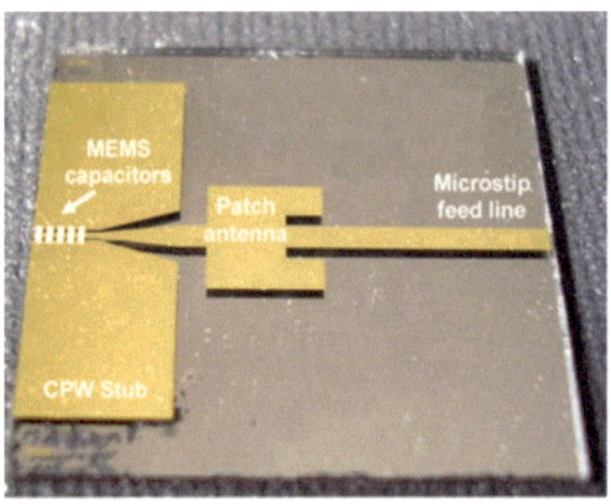

Figure 12. Frequency tunable microstrip patch antenna integrating MEMS capacitors [14]

When a DC voltage is applied, the height of the MEMS bridges on the stub is varying, and thus the loading capacitance is also changing. Therefore, as showed Figure 13 the matching frequency around 16.05 GHz shifts down to 15.75 GHz as the actuation voltage is increased from 0 to 11.9 V, where the height of the capacitive gap changes from 1.5 μm to 1.4 μm.

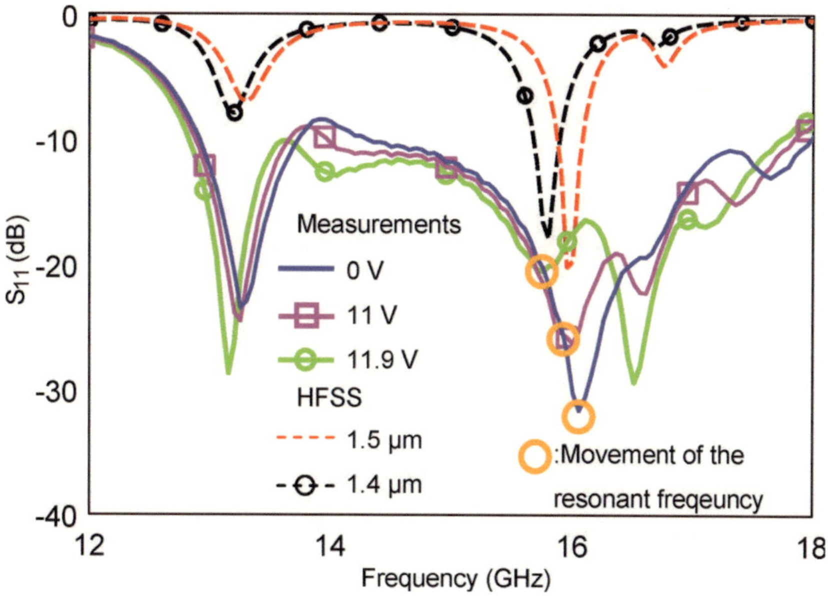

Figure 13. $|S_{11}|$parameters for different actuation voltages and simulation results [14]

Discrete frequency tuning can be illustrated with a reconfigurable annular slot antenna with a monolithic integration of MEMS actuators presented by B.A. Cetiner in [15]. The architecture and a photograph of the microstrip-fed reconfigurable antenna annular slot are shown in Figure 14. The antenna has two concentric circular slots. According to MEMS switch S_1 state, they can be individually excited in order to achieve frequency reconfigurability. S_2 and S_3 switches enable the metallic annular ring, which stays between the outer and inner slots, to be shorted to RF ground.

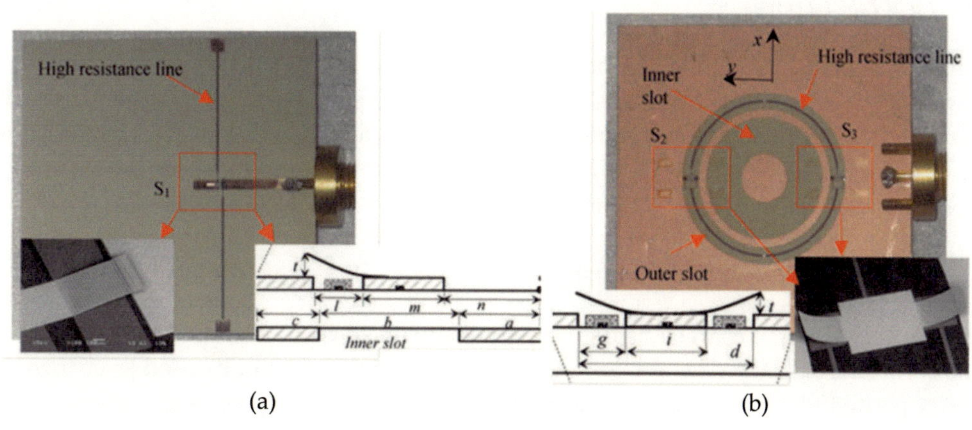

Figure 14. Microstrip feeding line integrating a single-arm MEMS switch (a) and the annular slot integrating two double-arm MEMS actuators [15]

The measured $|S_{11}|$ parameters (Figure 15) show that when MEMS switches are activated (down-state) by applying DC bias voltages, the antenna working band is around 5.2 GHz. Vice-versa, when MEMS switches are in the up-state, the antenna is working at 2.4 GHz.

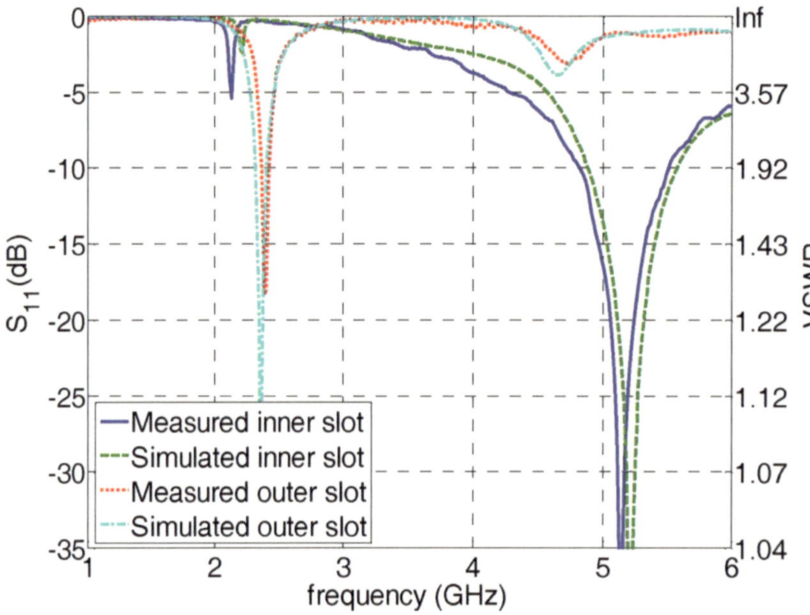

Figure 15. Measured and simulated $|S_{11}|$parameters for MEMS switches activated (5.2 GHz) and deactivated (2.4 GHz) [15]

2.1.4. Field Effect Transistor

Continuous frequency tuning can be achieved by using Field Effect Transistor. In [16], S. Kawasaki presents a slot antenna loaded with two one-port reactive FET. The electrically length of the slot is changing according to the voltage bias applying on the FET. The measured $|S_{11}|$ parameters (see Figure 16) show a 10% frequency tuning range for a gate tuning voltage (V_{gs}) going from 0V to-0.6V while the drain voltage (V_{ds}) is tuned from 0V to 0.4V.

2.2. Agile antennas using tunable materials

Changing the material characteristics in a part of antenna designs also promise the ability to tune them in frequency. The application of a static electric field can be used to change the relative permittivity of a ferroelectric material or a liquid crystal, respectively a static magnetic field can changed the relative permeability of a ferrite. In case of printed antennas, these changes modify the effective electrical length of antennas, and then resulting in shifts of their operating frequencies.

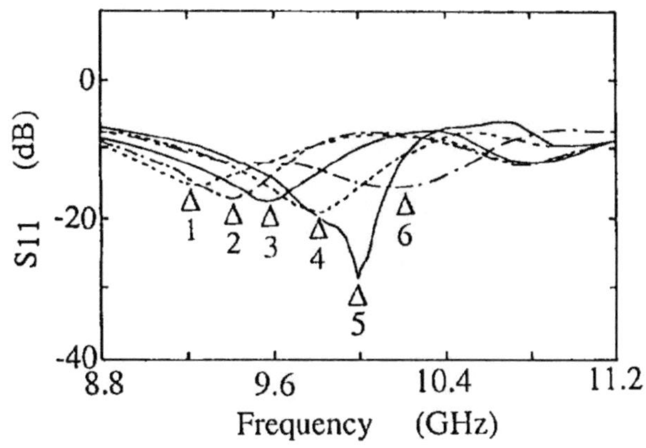

Figure 16. Measured $|S_{11}|$parameters for both gate and drain tuning voltages [16]

2.2.1. Ferroelectric materials

Lead-based perovskite ceramics such as $PbZr_xTi_{1-x}O_3$ (PZT) have been the leaders, for the past 50 years, on ferroelectric material research [17] for electronic devices, sensors, actuators, and medical ultrasonic transducers, owing to their good dielectric properties over a wide temperature range. Due to health care and environmental regulations, restriction of hazardous substances as lead has been required [18]. Since the last 10 years, many efforts have been mainly devoted in the field of microwave applications to $Ba_xSr_{1-x}TiO_3$ (BST) material which is one of the most attractive materials [19] because it presents high dielectric constant, relatively low dielectric loss, interesting tunability and small temperature dependence. In fact, in BST, the Curie temperature (Tc) which defines the ferroelectric/paraelectric transition is tuned by controlling the Ba/Sr ratio. More recently some other ferroelectric ceramics such as the tantalate niobate oxide $KTa_xNb_{1-x}O_3$ (KTN) or the sodium bismuth titanate $Na_{0.5}Bi_{0.5}TiO_3$ (BNT) and its solid solutions BNT-BT are emerging.

Two different methods exist to polarize a ferroelectric material with a static electric field. A better tunability is obtained when the static electric field is perpendicular to the two electrodes. For antenna application point of view, antenna designs integrating ferroelectric materials have to move toward this kind of polarization in order to have a better reconfigurability. However, many efforts have to be devoted from realization and also simulation point of views. Therefore, compact antenna community exhibits only few papers of this kind of antennas.

V. K. Palukuru et al. [20] present a tunable antenna using an integrated ferroelectric-thick film made of BST material. The antenna is depicted in Figure 17. It exhibits a folded slot antenna loading with a BST thin film varactor. In order to tune the dielectric permittivity of the BST film, a DC bias voltage is applied thanks to a bias-T component, which was attached to the Vector Network Analyzer. The BST varactor is placed over the radiating slot: the upper electrode (0.2 mm 0.2 mm) is part of the antenna's metallization and the lower electrode is the antenna's ground plane. In order to reduce the capacitance of the varactor, a slight horizontal offset is used between the electrodes. Therefore, the electric field for biasing the material is both in the vertical and the horizontal directions.

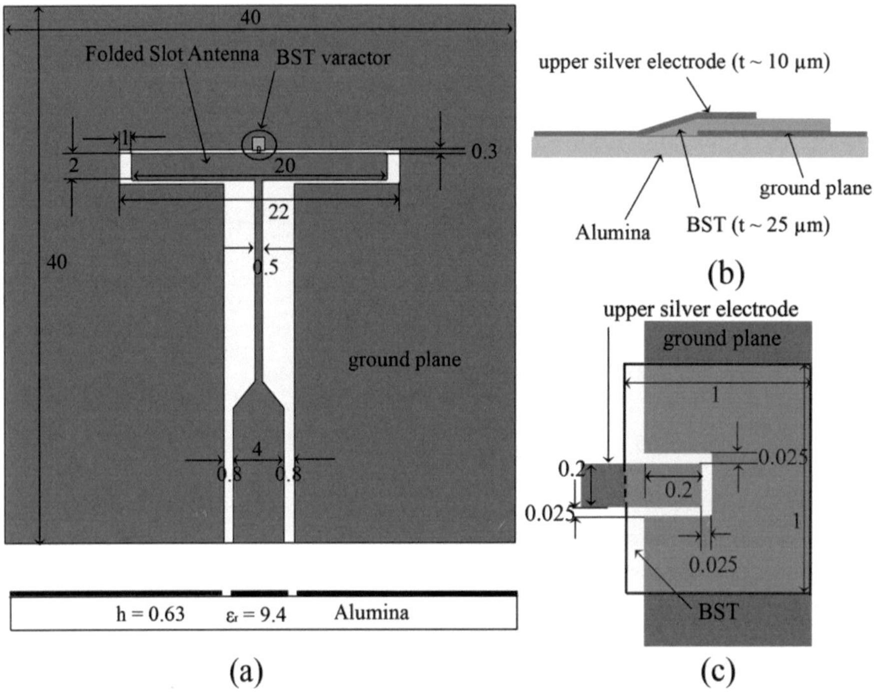

Figure 17. Folded slot antenna with the BST varactor (a), side cross-section (b) and top view (c) of the BST varactor [20]

The $|S_{11}|$ parameters (Figure 18) show that a frequency tunability of 3.5% can be obtained with a change of the bias voltage from 0V to 200V.

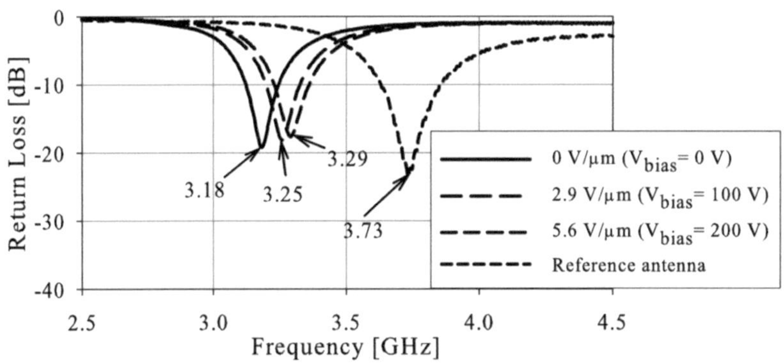

Figure 18. Measured $|S_{11}|$parameters for different DC-bias voltages [20]

H. Jiang presents a coplanar waveguide (CPW) square-ring slot antenna as showed Figure 19 [21]. Nine shunt ferroelectric BST thin film varactors are integrated with the CPW antenna structure achieving both antenna miniaturization and reconfiguration.

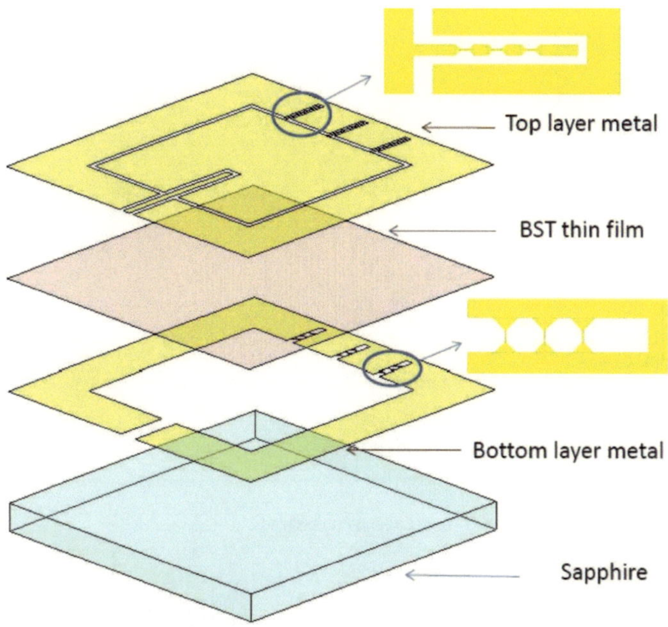

Figure 19. Coplanar waveguide square-ring slot antenna integrating BST material [21]

Figure 20 shows the measured $|S_{11}|$ parameter with DC bias voltages from 0 V to 7 V. Therefore the antenna working band is continuously tuned from 5.28 GHz up to 5.77 GHz.

2.2.2. Liquid crystal

Another technological approach for designing an agile antenna is the use of liquid crystal as a tunable dielectric. Indeed, the characterizations of liquid crystals [22-24] have shown that they are promising tunable materials for microwave applications, especially for operating frequencies above 10 GHz. The material features low dielectric loss and continuous tunability with low bias power consumption. The literature show that some microwave applications are using liquid crystals, e.g. for polarization agile antenna [25], tunable patch antennas [26], reflectarrays [27-28], filters [29], resonators [30] and variable delay lines [31-33].

In this framework, L. Liu presents in [34] a tunable patch antenna using a liquid crystal. Its operating frequency is around 5 GHz with a tuning range around 4% in measurement. The Figure 21 presents the antenna geometry composed of three layers of Taconic substrate. The liquid crystal is injected in the middle layer just under the microstrip patch and between the ground plane and patch top hat. A DC bias voltage was applied between the patch and ground across the liquid crystal using a bias tee at the feed input.

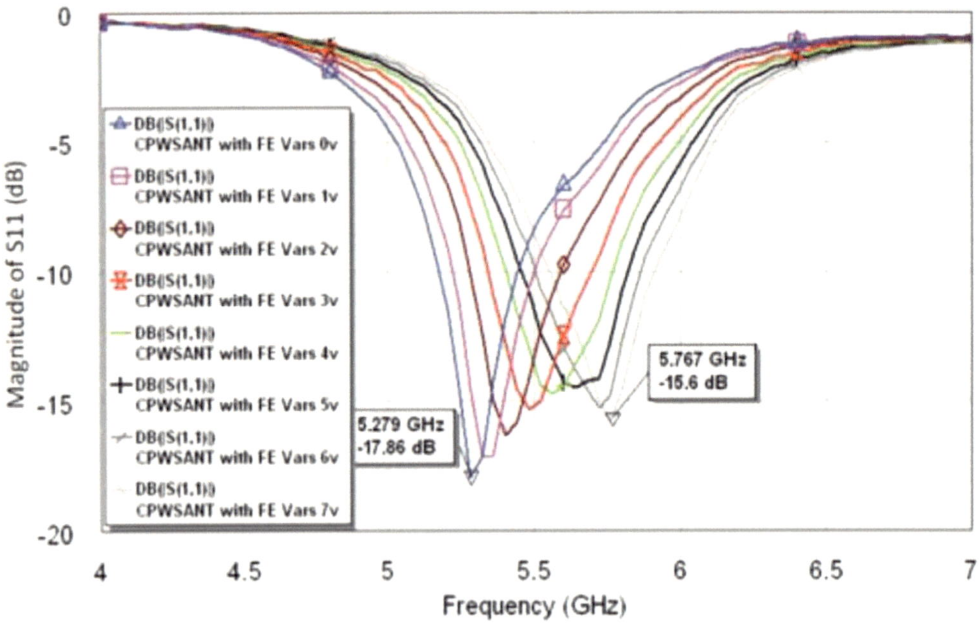

Figure 20. Measured $|S_{11}|$parameters for different DC-bias voltages [21]

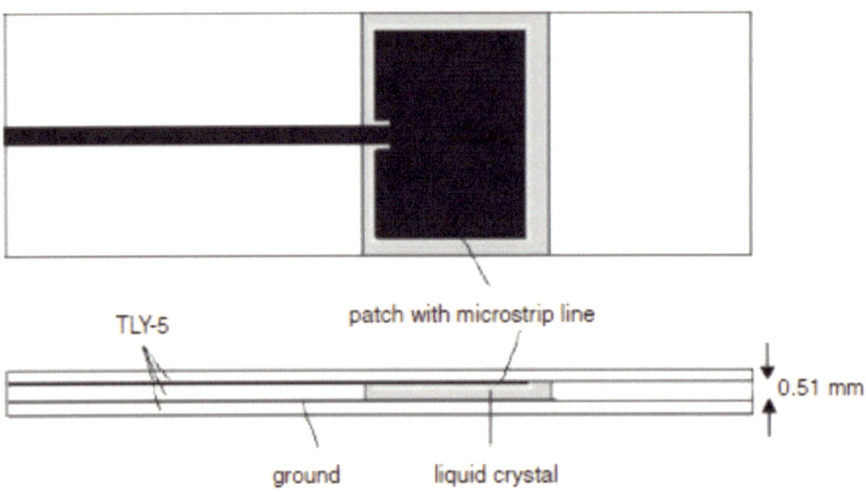

Figure 21. Tunable patch antenna using a liquid crystal [34]

The Figure 22 shows the measured return losses for the patch antenna for three states of liquid crystal bias: 0V, 5V and 10V. The 0V state reveals that the used liquid crystal without DC bias presents somewhat high losses around 0.12 for the loss tangent. Thus a 4% frequency tuning can be achieved with relatively poor radiation efficiencies, i.e. 14% at least (at 0V) to 40% (at 10V).

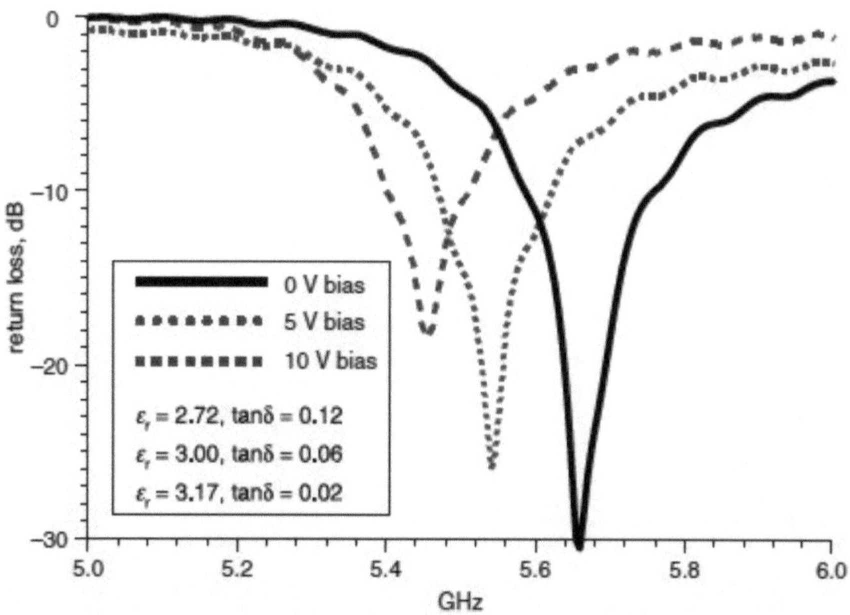

Figure 22. Measured $|S_{11}|$parameters for different DC-bias voltages [34]

2.2.3. Ferrite materials

Frequency-tuned ferrite-based antennas are rarely presented in the literature. In [35] and [36], authors study patch antennas on ferrite substrates (Figure 23) whereas A. Petosa presents in [37] a ferrite resonator antenna. In this latter, biasing of the ferrite with a static magnetic field is achieved using a permanent magnet. The magnet was located under the ferrite antenna beneath the ground plane. For a parallel magnetic-bias orientation, the resonance frequency can be tuned on 8% of bandwidth. That scales to 9% for a perpendicular magnetic-bias orientation.

All results presented in the literature and investigated the properties of ferrite-based microstrip antennas indicate that factors including non-uniform bias fields and the multiple modal field distributions excited in a bulk ferrite substrate may preclude their use in practical applications.

Now that an overview of the most common used techniques for compact antennas to become active has been presented, the next part will address both challenges and how to integrate active components in order to maximize the antenna performances and efficiency.

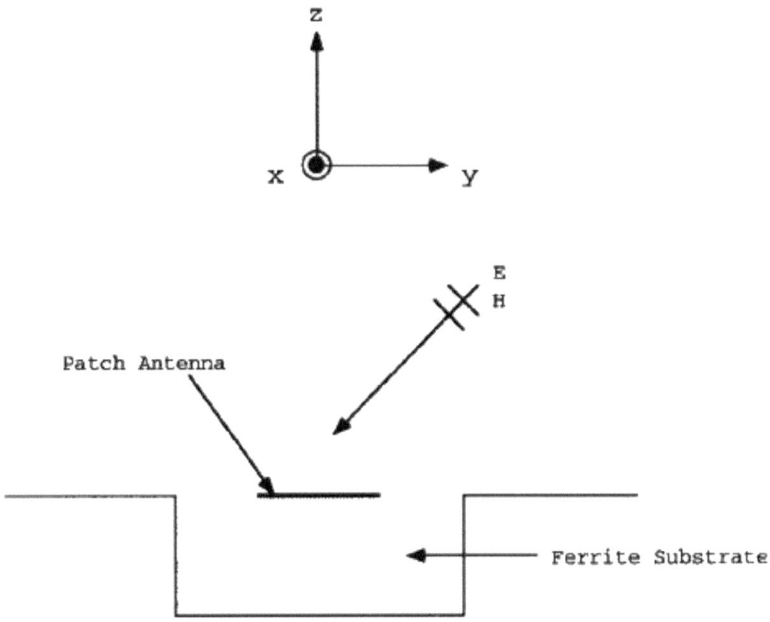

Figure 23. Patch antenna on a ferrite substrate [37]

3. Relevant parameter for frequency agile antenna studies – New approach for wireless applications.

3.1. How to integrate an active component – Challenges

To implement the reconfigurability in an antenna, the knowledge of the active component or the tunable material is essential. Even in case of commercial components, as varactor or PIN diodes, users and particularly the microwave community do not have enough parameters and information at RF frequencies. Some papers detail the integration of the RF equivalent circuit of the used component [10],[13].

The varactor diodes integrated in an antenna for frequency reconfigurability is the most popular way. Thus, this section will focus on varactor diodes issues. However, arguments can be extended to other frequency tuning methods.

In [10], the paper completes the lack of information related to most of varactor diode datasheets. Indeed, constructor only provide characteristics at low frequencies and do not give enough parameters for antenna application point of view, such as capacitance values, serial resistance and accepted power at RF frequencies. The chosen varactor in this paper has been characterized according to antenna designer criteria and its electromagnetic model has been deduced. This example is chosen in the framework of this chapter.

To correctly explain this example, the next subsection will investigate the place where the varactor can be integrated. Following the presentation of the varactor diode manufacturer's datasheet, a complete characterization meeting antenna designer's criteria will be explained. That will lead to the determination of the varactor diode S parameters. With the knowledge of the latter, two methods will be explained and described to reach an accurate antenna simulation and realization:

- The electromagnetic equivalent circuit of the varactor diode can be deduced from the S parameters thanks to Agilent ADS.
- The S parameters of the varactor diode can be directly injected in the electromagnetic simulator CST Microwave Studio® and the antenna performances deduced thanks to a co-simulation.

3.1.1. Varactor diode area in an antenna

The varactor diode is a tunable capacitor which loads the antenna in order to artificially increase its electrical length. To be the most efficient, it must be placed where the electrical field is maximum. Take a folded Inverted F Antenna for example presented Figure 24. The maximum of the electrical field is at the end of the radiating element (see Figure 24). To be integrated at this place and joined the ground plane at the same time, the varactor diode can be soldered between the ribbon and the ground. In most cases, DC-block capacitors have to be added for the varactor's DC-bias not to be shunt.

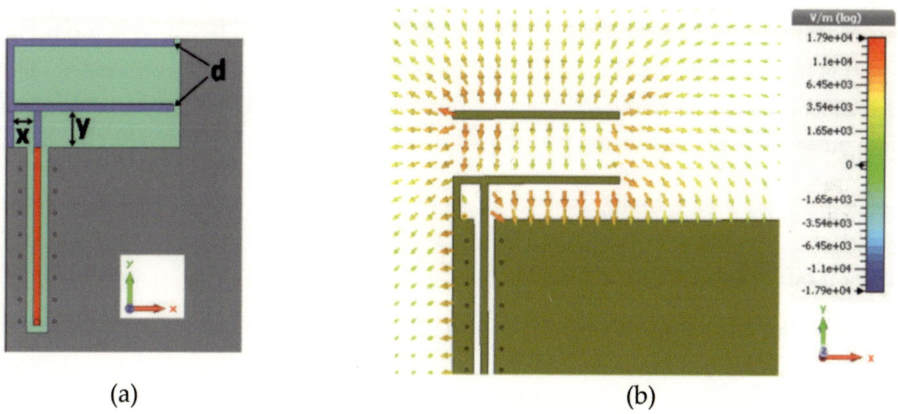

Figure 24. Inverted F Antenna Measured (a) and the total electric field on the radiating element (b)

3.1.2. Varactor diode datasheet

In [10], the chosen GaAs hyperabrupt varactor diode MGV125-22 (Aeroflex Metelics) [38] has a capacity range between 0.2 pF and 2 pF for 0 to 22 Volts tuning voltage as shown Figure 25. However, these values are given as a rough line and the datasheet does not give enough parameters for high frequencies antenna application's point of view. Indeed, the values for junction capacitance C_j (Figure 25) and the quality factor Q are supplied by the manufacturer and are almost always specified at a low frequency. The Figure 25 shows the widely used varactor diode model with L_p and C_p the values of the package inductance and capacitance.

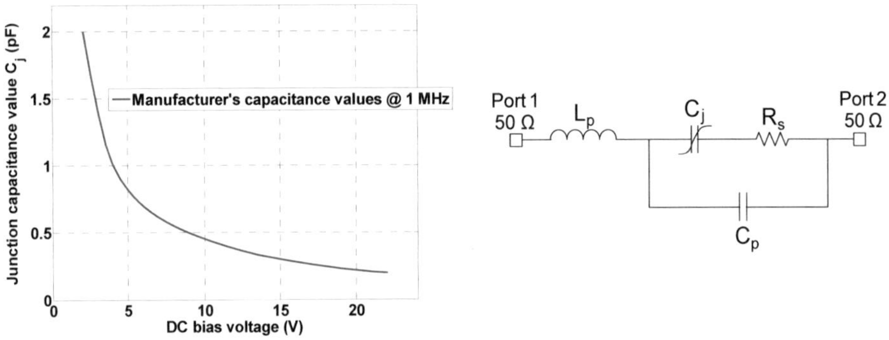

(a) (b)

Figure 25. Manufacturer's capacitance value (extracted at 1MHz) versus the DC bias voltage (a) and the Varactor diode model (b)

In the MGV125-22 case, junction capacitance values are specified at 1 MHz and the Q factor equals 3000 at 50 MHz for a DC bias voltage of –4 Volts. Q is defined by $Q=1/(\omega C_j R_s)$. This formula can be used to calculate the series resistance R_s of the varactor model at the measured frequency, its value is assumed to be constant with reverse voltage. Thus, at 50 MHz $R_s=1.06\Omega$. It is important to note that R_s impacts directly the antenna total efficiency. A too high value (from 3Ω) is basically penalizing for antenna performances. This enhances the need to assess its value at microwave frequencies. For this purpose, the hyperabrupt varactor diode has to be characterized close to operating conditions (here between 470 MHz and 862 MHz).

3.1.3. Varactor diode characterization

- First method: Electromagnetic model

The varactor diode is soldered on a 50 Ω impedance microstrip line as shown Figure 26. A dedicated TRL (Through-Reflect-Line) calibration kit is manufactured (Figure 26) in order to de-embed both connectors and lines. Thus S parameters of the single varactor diode can be deduced. Considering the varactor diode model previously presented in Figure 26, C_j, L_p, C_p and R_s values can be deduced for each voltage and for a constant injected power of-10 dBm. Therefore, model component values are adjusted (see Figure 26) in order their S parameters to correspond with the measured ones. The Figure 26 shows the comparison between S parameters of the determined electromagnetic model (Agilent ADS) and the measured ones for 2 Volts and 10 Volts DC bias voltages.

These results are given as an example and the same work has been done for varactor reverse bias voltages varying from 2V to 22V with a 2V step. As expected, the corresponding electromagnetic model presents constant values according to the DC bias voltage (Figure 27):

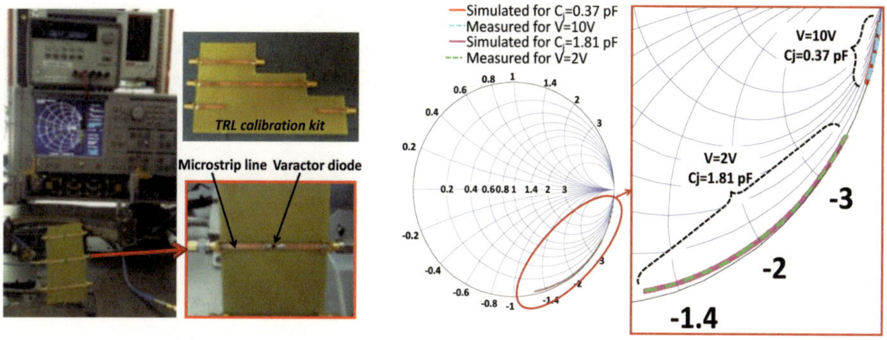

(a) (b)

Figure 26. Characterization of the varactor diode (a) and the Comparison between the electromagnetic model and the measurement on S_{11} parameter on the [400 MHz – 1 GHz] frequency band (b)

L_p=3.821nH, C_p=0.08pF and R_s=1.8Ω. The C_j value presented in Figure 27 decreases as a function of DC bias voltage value.

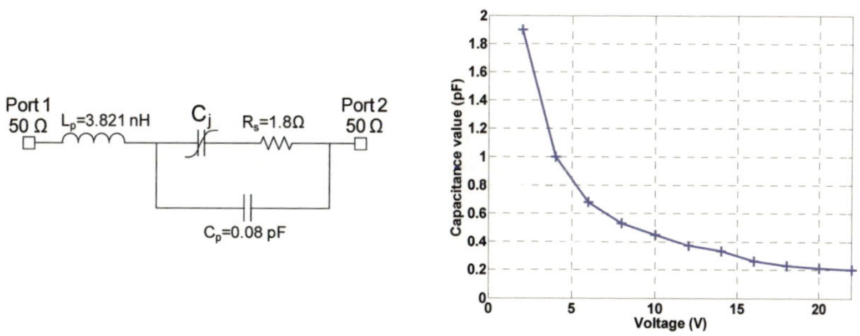

(a) (b)

Figure 27. Electromagnetic model with the capacitance values C_j versus the DC bias voltage (b)

- Second method: Co-simulation

Another way is to directly insert the S parameters touchstone file of the varactor diode in the antenna electromagnetic simulation as illustrated in Figure 28.

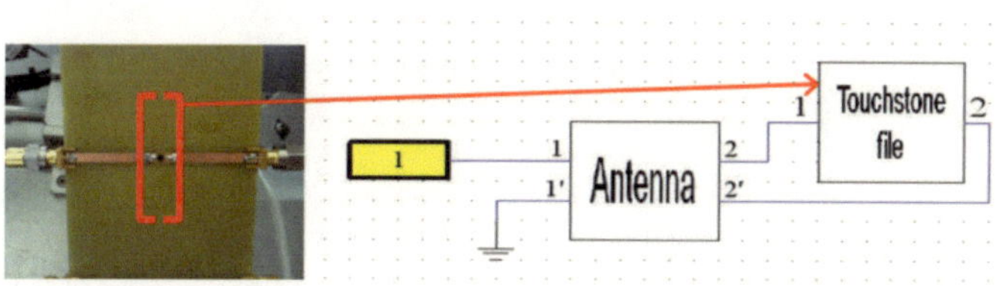

Figure 28. Co-simulation of the antenna including measured S parameters of the diode

Thanks to the previous TRL calibration, only the varactor diode S parameters are inserted in the simulator. By this way, both antenna and varactor diode are combined and the S parameters of the global device can be directly simulated. An example (presented paragraph 3.3) will confirm that the two previous methods exhibit similar antenna performances.

3.2. Limitations of currently varactor diode method – Power characterization

Figure 29 provides some information regarding the accepted power by the varactor diode: high injected power involves some varactor diode distortions.

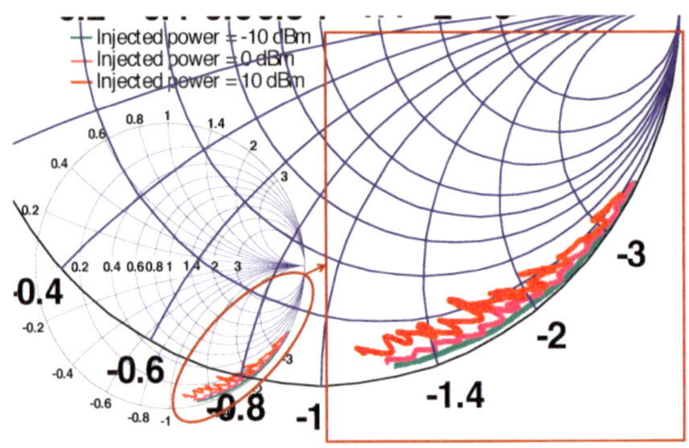

Figure 29. Measured S parameters according to the injected power

This figure presents S parameters of the diode for only three values of injected power:-10 dBm, 0 dBm and 10 dBm. The non-linear distortion of the diode has been studied. It reveals that the varactor diode model well fits measurements for an injected RF power lower than-5 dBm. Beyond this injected power (see for 0 dBm), no varactor model can fit the measurement. Regarding antenna's parameters, the following example will show that a large RF power (upper than-5 dBm) involves a mismatched antenna. As far as the DVB-H application, the system is only working in receiving mode, the diode distortion will never appear and the linear electromagnetic model can be used and integrated in the electromagnetic simulator. Indeed, for receiver devices, the antenna accepted power is far lower than – 5 dBm.

3.3. Example of a basic tunable DVB-H antenna

This subsection investigates an example to show the interest of the previous varactor diode characterization. This was briefly presented in [4], it is completed here by adding the first method (electromagnetic model) and the power characterization. A basic IFA prototype loaded by the same varactor diode (see Figure 30) has been manufactured.

It has been measured for a-10 dBm RF power and its performances compared with three simulations: with both presented methods and with the varactor diode's datasheet (Figure 31). $|S_{11}|$ parameters show that both investigated methods and measurement present a good agreement. Moreover, they are different from $|S_{11}|$ parameters determined with the varactor diode datasheet. That underlines the relevance of the varactor diode characterization. Regard-

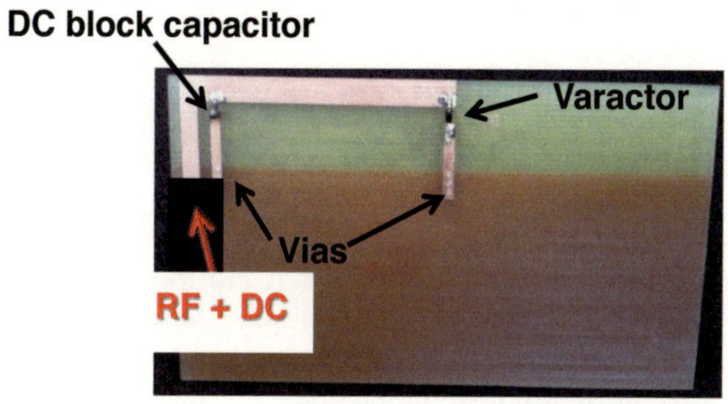

Figure 30. Basic IFA prototype

ing the antenna total efficiency, it equals 50% in the real case whereas it reaches 60% when the varactor datasheet is used in electromagnetic simulations.

Power characterization is illustrated on Figure 31. This figure reminds the measured $|S_{11}|$ parameter for a power of-10 dBm. For a 10 dBm injected power, the measured $|S_{11}|$ parameter is compared with the simulated one with the second method.

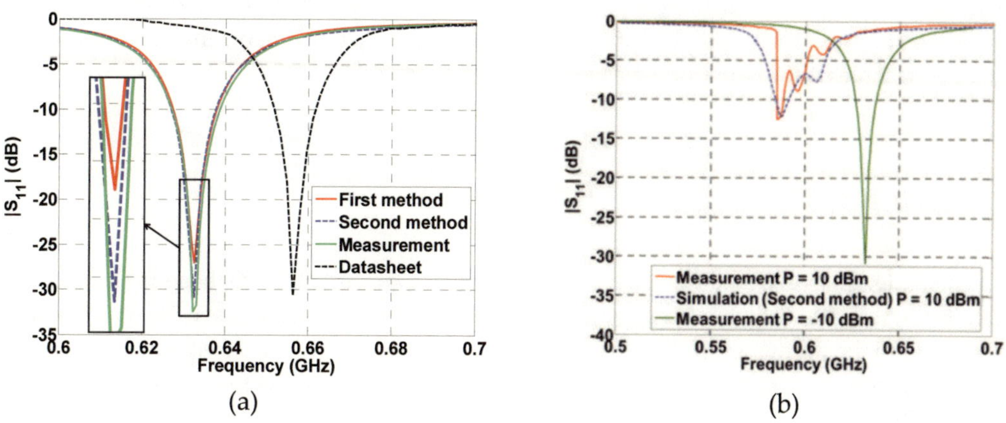

Figure 31. $|S_{11}|$ parameters for the different methods (a) and according to the injected power (b)

There is a good agreement between the measurement and the simulation. This figure shows that the non-linearity of the varactor diode disturbs the $|S_{11}|$ parameter of the antenna. Thus, this kind of varactor diode has to be used only for reception devices.

Therefore, this section has presented how to integrate and characterize a varactor diode. It reveals the importance of the diode characterization in a design flow dedicated to antenna structure.

3.4. Discussion and trade-offs between agility techniques and physical limitations

According to the aimed application, trade-offs are necessary to design a frequency tunable antenna.

- For discrete frequency tuning, PIN diodes or MEMS switches can be planed.
- For continuous frequency tuning, which is often aiming for compact antennas, varactor diodes, MEMS variable capacitor and tunable materials can be used.

However, previous paragraphs have revealed that varactor diodes are not usable for transmitter devices because of their non-linearity for considering power levels.

The RF characterization of ferroelectric films shows high power handling capability [39]. Good permittivity tunability may be obtained if both materials properties and variable capacitor sizes have been properly dimensioned. The conclusion is the same for MEMS variable capacitor.

Studies on both ferroelectric material and MEMS capacitor merit extended investigations because these solutions seem to be the best and most promising alternatives faced with varactor diodes.

Performing rigorous full-wave analysis of these new components is the new challenge to extract their accurate electromagnetic models. Antennas would be optimized by considering the real response of these components. These investigations would enable the co-development of antennas integrating components' electromagnetic models.

4. Conclusion

To conclude, an overview of compact and frequency agile antenna has been presented and detailed in this chapter while mentioning a lot of literature references. A special part has been dedicated to the presentation of the most common method to achieve a frequency tuning: the use of varactor diodes. Their integration within an antenna to be the most efficient has been shown. The study exhibits varactor diodes characterization and also reveals their limitation. A summary of the presented methods according to the intended application has been presented. Eventually, some ideas on varator diodes alternatives have been proposed in order to make antenna tunability viable for transmitter devices.

Author details

L. Huitema and T. Monediere

University of Limoges, Xlim Laboratory, France

References

[1] Huitema, L.; Sufyar, S.; Delaveaud, C.; D'Errico, R., "Miniature antenna effect on the ear-to-ear radio channel characteristics," *Antennas and Propagation (EUCAP), 2012 6th European Conference on*, vol., no., pp.3402,3406, 26-30 March 2012

[2] Libo Huang; et al., "Electrically Tunable Antenna Design Procedure for Mobile Applications," *Microwave Theory and Techniques, IEEE Transactions on*, vol.56, pp. 2789-2797, Dec. 2008

[3] Li, Y., Zhang, Z., Chen, W., Feng, Z. and Iskander, M. F. (2010), A compact DVB-H antenna with varactor-tuned matching circuit. Microw. Opt. Technol. Lett., 52: 1786–1789

[4] Huitema, L.; Reveyrand, T. et al., "A compact and reconfigurable DVB-H antenna for mobile handheld devices," *Antennas and Propagation (EUCAP), Proceedings of the 5th European Conference on*, vol., no., pp.1314-1317, 11-15 April 2011

[5] Canneva, F.; Ribero, J.; Staraj, R., "Tunable antenna for DVB-H band," *Antennas and Propagation (EuCAP), 2010 Proceedings of the Fourth European Conference on*, vol., no., pp.1,3, 12-16 April 2010

[6] Bhartia, P.; Bahl, I.J., "A frequency agile microstrip antenna," *Antennas and Propagation Society International Symposium, 1982*, vol.20, no., pp.304,307, May 1982

[7] Erfani, E.; Nourinia, J.; Ghobadi, C.; Niroo-Jazi, M.; Denidni, T.A., "Design and Implementation of an Integrated UWB/Reconfigurable-Slot Antenna for Cognitive Radio Applications," *Antennas and Wireless Propagation Letters, IEEE*, vol.11, no., pp. 77,80, 2012

[8] Behdad, N.; Sarabandi, K., "A varactor-tuned dual-band slot antenna," *Antennas and Propagation, IEEE Transactions on*, vol.54, no.2, pp.401,408, Feb. 2006

[9] Behdad, N.; Sarabandi, K., "Dual-band reconfigurable antenna with a very wide tunability range," *Antennas and Propagation, IEEE Transactions on*, vol.54, no.2, pp.409,416, Feb. 2006

[10] Huitema, L.; Reveyrand, T.; Mattei, J.-L.; Arnaud, E.; Decroze, C.; Monediere, T., "Frequency Tunable Antenna Using a Magneto-Dielectric Material for DVB-H Application," *Antennas and Propagation, IEEE Transactions on*, vol.61, no.9, pp.4456,4466, Sept. 2013

[11] Laheurte, J.-M., "Switchable CPW-fed slot antenna for multifrequency operation," *Electronics Letters*, vol.37, no.25, pp.1498,1500, 6 Dec 2001

[12] Outi Kivekäs, Jani Ollikainen, and Pertti Vainikainen, "Frequency-tunable internal antenna for mobile phones", 12th International Symposium on Antennas (JINA 2002). Nice, France, 12-14 November 2002, volume 2, pages 53-56

[13] Peroulis, D.; Sarabandi, K.; Katehi, L. P B, "Design of reconfigurable slot antennas," *Antennas and Propagation, IEEE Transactions on*, vol.53, no.2, pp.645,654, Feb. 2005

[14] Erdil, Emre; Topalli, K.; Unlu, M.; Civi, O.A.; Akin, T., "Frequency Tunable Microstrip Patch Antenna Using RF MEMS Technology," *Antennas and Propagation, IEEE Transactions on*, vol.55, no.4, pp.1193,1196, April 2007

[15] Cetiner, B.A.; Crusats, G.R.; Jofre, L.; Biyikli, Necmi, "RF MEMS Integrated Frequency Reconfigurable Annular Slot Antenna," *Antennas and Propagation, IEEE Transactions on*, vol.58, no.3, pp.626,632, March 2010

[16] Kawasaki, S.; Itoh, T., "A slot antenna with electronically tunable length," *Antennas and Propagation Society International Symposium, 1991. AP-S. Digest*, vol., no., pp. 130,133 vol.1, 24-28 June 1991

[17] N. Setter, D. Damjanovic, et al., Ferroelectric thin films: Review of materials, properties, and applications, J. Appl. Phys. 100, (2006)

[18] P. K. Panda, Review: environmental friendly lead-free piezoelectric materials, J. Mater. Sci. 44, 5049–5062, (2009)

[19] A. K.Tagantsev, V. O. Sherman, K. F. Afanasiev, J. Venkatesh and N. Setter, Ferroelectric Materials for Microwave Tunable Applications, J. Electroceram., 11, 5-66, 2003

[20] Palukuru, V.K.; Komulainen, M.; Tick, T.; Perantie, J.; Jantunen, Heli, "Low-Sintering-Temperature Ferroelectric-Thick Films: RF Properties and an Application in a Frequency-Tunable Folded Slot Antenna," *Antennas and Wireless Propagation Letters, IEEE*, vol.7, no., pp.461,464, 2008

[21] Hai Jiang; Patterson, M.; Brown, D.; Chenhao Zhang; KuanChang Pan; Subramanyam, G.; Kuhl, D.; Leedy, K.; Cerny, C., "Miniaturized and Reconfigurable CPW Square-Ring Slot Antenna Loaded With Ferroelectric BST Thin Film Varactors," *Antennas and Propagation, IEEE Transactions on*, vol.60, no.7, pp.3111,3119, July 2012

[22] Schaub, D.E.; Oliver, D.R., "A Circular Patch Resonator for the Measurement of Microwave Permittivity of Nematic Liquid Crystal," *Microwave Theory and Techniques, IEEE Transactions on*, vol.59, no.7, pp.1855,1862, July 2011

[23] Mueller, S.; Penirschke, A.; Damm, C.; Scheele, P.; Wittek, M.; Weil, Carsten; Jakoby, R., "Broad-band microwave characterization of liquid crystals using a temperature-controlled coaxial transmission line," *Microwave Theory and Techniques, IEEE Transactions on*, vol.53, no.6, pp.1937,1945, June 2005

[24] Bulja, S.; Mirshekar-Syahkal, D.; James, R.; Day, S.E.; Fernandez, F.A., "Measurement of Dielectric Properties of Nematic Liquid Crystals at Millimeter Wavelength," *Microwave Theory and Techniques, IEEE Transactions on*, vol.58, no.12, pp.3493,3501, Dec. 2010

[25] Karabey, O.H.; Bildik, S.; Bausch, S.; Strunck, S.; Gaebler, A.; Jakoby, R., "Continuously Polarization Agile Antenna by Using Liquid Crystal-Based Tunable Variable Delay Lines," *Antennas and Propagation, IEEE Transactions on*, vol.61, no.1, pp.70,76, Jan. 2013

[26] Luyi Liu; Langley, R., "Electrically small antenna tuning techniques," *Antennas & Propagation Conference, 2009. LAPC 2009. Loughborough*, vol., no., pp.313,316, 16-17 Nov. 2009

[27] Wenfei Hu; Arrebola, M.; Cahill, R.; Encinar, J.A.; Fusco, V.; Gamble, H.S.; Alvarez, Y.; Las-Heras, F., "94 GHz Dual-Reflector Antenna With Reflectarray Subreflector," *Antennas and Propagation, IEEE Transactions on*, vol.57, no.10, pp.3043,3050, Oct. 2009

[28] Yazdanpanahi, M.; Bulja, S.; Mirshekar-Syahkal, D.; James, R.; Day, S.E.; Fernandez, F.A., "Liquid-crystal-based mm-wave tunable resonator," *Microwave Conference (EuMC), 2010 European*, vol., no., pp.1233,1236, 28-30 Sept. 2010S.P.; Grant, N., "Phase agile reflectarray cells based on liquid crystals," *Microwaves, Antennas & Propagation, IET*, vol.1, no.4, pp.809,814, Aug. 2007

[29] Goelden, F.; Gaebler, A.; Karabey, O.; Goebel, M.; Manabe, A.; Jakoby, R., "Tunable band-pass filter based on Liquid Crystal," *German Microwave Conference, 2010*, vol., no., pp.98,101, 15-17 March 2010

[30] Yazdanpanahi, M.; Bulja, S.; Mirshekar-Syahkal, D.; James, R.; Day, S.E.; Fernandez, F.A., "Liquid-crystal-based mm-wave tunable resonator," *Microwave Conference (EuMC), 2010 European*, vol., no., pp.1233,1236, 28-30 Sept. 2010

[31] Kuki, T.; Fujikake, H.; Nomoto, T., "Microwave variable delay line using dual-frequency switching-mode liquid crystal," *Microwave Theory and Techniques, IEEE Transactions on*, vol.50, no.11, pp.2604,2609, Nov 2002

[32] Goelden, F.; Gaebler, A.; Goebel, M.; Manabe, A.; Mueller, S.; Jakoby, R., "Tunable liquid crystal phase shifter for microwave frequencies," *Electronics Letters*, vol.45, no. 13, pp.686,687, June 18 2009

[33] Karabey, O.H.; Goelden, F.; Gaebler, A.; Strunck, S.; Jakoby, R., "Tunable loaded line phase shifters for microwave applications," *Microwave Symposium Digest (MTT), 2011 IEEE MTT-S International*, vol., no., pp.1,1, 5-10 June 2011

[34] Liu, L.; Langley, R.J., "Liquid crystal tunable microstrip patch antenna," *Electronics Letters*, vol.44, no.20, pp.1179,1180, September 25 2008

[35] Brown, A.D.; Volakis, J.L.; Kempel, Leo C.; Botros, Y.Y., "Patch antennas on ferromagnetic substrates," *Antennas and Propagation, IEEE Transactions on*, vol.47, no.1, pp. 26,32, Jan 1999

[36] Mishra, R.K.; Pattnaik, S.S.; Das, N., "Tuning of microstrip antenna on ferrite substrate," *Antennas and Propagation, IEEE Transactions on*, vol.41, no.2, pp.230,233, Feb 1993

[37] Petosa, A.; Mongia, R.K.; Cuhaci, M.; Wight, J.S., "Magnetically tunable ferrite resonator antenna," *Electronics Letters*, vol.30, no.13, pp.1021,1022, 23 Jun 1994

[38] *Aeroflex Metelics datasheet*, Available: http://www.aeroflex.com/AMS/Metelics/pdfiles/MGV_Series_Hyperabrupt_A17041.pdf

[39] Laur, V.; Costes, R.; Houndonougbo, F.; et al., "Microwave study of tunable planar capacitors using mn-doped $ba_{0.6}sr_{0.4}tio_3$ ceramics," *Ultrasonics, Ferroelectrics and Frequency Control, IEEE Transactions on*, vol.56, no.11, pp.2363-2369, November 2009

Chapter 5

Active Compact Antenna for Broadband Applications

Y. Taachouche, M. Abdallah, F. Colombel,
G. Le Ray and M. Himdi

Additional information is available at the end of the chapter

http://dx.doi.org/10.5772/58839

1. Introduction

The recent development of wireless communication technology and the miniaturization of electronics components increase the demand for compact systems applications including small antennas. One of the major challenges is the integration of antennas inside devices in a limited area. The main characteristics of these antennas are large frequency bandwidth and small size.

Many passive antennas such as monopole, dipole and printed antenna have been largely studied to yield small size relatively to the wavelength or broadband behavior. Previous studies have shown that antenna miniaturization impacts negatively antenna bandwidth and impedance matching [1].

The active antennas have found a wide interest for industrial applications in last years. The terminology of the active antenna indicates that the passive antenna elements are combined with an active device on the same substrate to provide a non-separated device and to improve antenna performances, especially in the field of size reduction and frequency bandwidth. The ability to adjust the size reduction and the frequency bandwidth of an active antenna is also very suitable when the antenna is included inside devices with many components located in a limited area.

In this chapter, we are interested in the improvements brought by the active antennas towards size reduction and the covered bandwidth. We present our work on compacts actives antennas in which we develop new techniques to reduce the size of antennas with good performances. Our main works is carried on two solutions; the first one is a broadband antenna with a very important size reduction. This solution corresponds to an active printed monopole integrating a bipolar transistor directly on the antenna structure without matching circuit. The second solution is a tunable narrow frequency band antenna operating on a wide band based on a printed loop antenna associated to a varactor diode.

2. Active monopole antenna

2.1. Introduction

Miniaturized and broadband antennas with ominidirectional coverage have attracted attention for industrial applications. For example, today's mobile phones are innovative devices that provide a wide variety of services to users. One of the most attractive mobile phone services is the entertainment services, and especially the functionality that allows users to listen to FM radios through their mobile phones. This development induces a growing demand of FM antennas for mobile phones, and the necessity of innovative technologies to develop internal small FM antennas, which replace the external wire antennas that exhibit current mobile phones in the market. For automobile application, many radio systems as communication system, FM radio reception are collocated on the top roof of the car, and the very low visual impact of the antenna is required. The goal of this section is to propose a compact antenna to replace the historical monopole to improve the integration capability of the antenna in VHF band applications systems.

The idea of using active antennas can be traced back to as early as 1928 [2]. A small antenna with electron tube was commonly used in radio broadcast receivers around 1MHz (Figure 1).

With the invention of high frequency transistor, the studies of active antennas have been performed in the years 60-70 [3-11].

Literature contribution on active antennas exhibits several advantages, the frequency bandwidth or signal to noise improvements compared to a passive antenna of the same size. In this contribution, we have started from the results given by Meinke [4] who have inserted the electronic components (tunnel diode and transistor) directly on the structure without matching network.

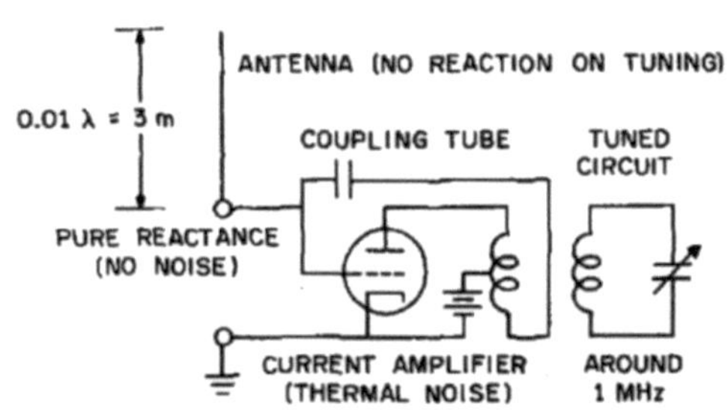

Figure 1. Active receiving antenna

A combination of a resonance half wavelength dipole with a VHF transistor has been proposed in [4]. The size of the antenna related to the wavelength is $\lambda/2$. As it is mentioned in [12], an active antenna provides new opportunities for many applications including the increase of bandwidth or the size reduction. In [13], a loop dipole has been proposed as a transmitting antenna. A total height of $\lambda/2000$ has been built and compared to a passive dipole which has the same height. The active antenna is very broadband and operates at very low frequencies besides its very small size. In [14], it has been explained that an active monopole fed by a microwave transistor provides a wider frequency bandwidth than a passive monopole.

In order to provide innovative designs, the growing interest for active antennas has required more accurate analysis method. A hybrid analysis including electromagnetic full wave and nonlinear circuit solver is used in [15] and accurate theoretical results validated with measurements on an oscillator active antenna are provided. In [16], the analysis method used in a circuit voltages generated by a CAD software as source distribution for a magnetic current radiation calculation to allow estimation of the integrated antenna is presented. In [17], theoretical or experimental methods are explained to study active antennas.

In this section, the problem of matching a short monopole antenna by including a transistor in the monopole structure is presented. We are interested in the influence of the transistor on the behavior of active antenna towards size reduction, bandwidth and gain. We investigate an active receiving antenna based on a printed monopole associated to a bipolar transistor. In the first part, we will present the antenna design working on the VHF/UHF bands and the theoretical approach. In the second part, we will exhibit results for two transistor configuration. Measurement results will be compared to the simulated ones and the influence of the transistor location on the monopole will be studied.

2.2. Design of the active monopole antenna

The structure of the active receiving antenna is shown in Figure 2. The antenna is a combination of a monopole and a high frequency bipolar transistor (BFR182). The antenna has been printed on a Neltec NX9300 substrate (ε_r=3, h=0.786 mm, tanδ=0.0023) and is placed above a limited square reflector plane (500×500×4 mm^3). As shown in Figure 2, the monopole has been cut in to two parts and the bipolar transistor BFR 182 is directly integrated between these two parts of the monopole without matching circuit. The antenna is connected to a SMA connector through a 50 Ω microstrip line [18-19].

We separate the printed structure into two zones. The first one is the area on which the monopole is printed; the second region (Area 2) welcomes the elements of the bias circuit of the transistor. A printed line, called "parasite" is added to connect the third pin of the transistor to ground in area 2 at a distance p from the monopole.

The active component used in our work is a bipolar PNP transistor BFR 182 used for low noise and high-gain broadband amplifier applications, and operates up to 8GHz. The transistor is located at h_2 over the ground plane and is used in two configurations. The first one is the common emitter configuration. In this case, the emitter of the transistor is grounded through the parasitic printed line, the transistor base and collector are respectively linked to the upper part and to the bottom part of the monopole.

The second one is the common collector configuration, the base of the transistor is connected to the upper part and the emitter of the transistor to the lower part of the monopole, the collector is connected to ground via the parasite line.

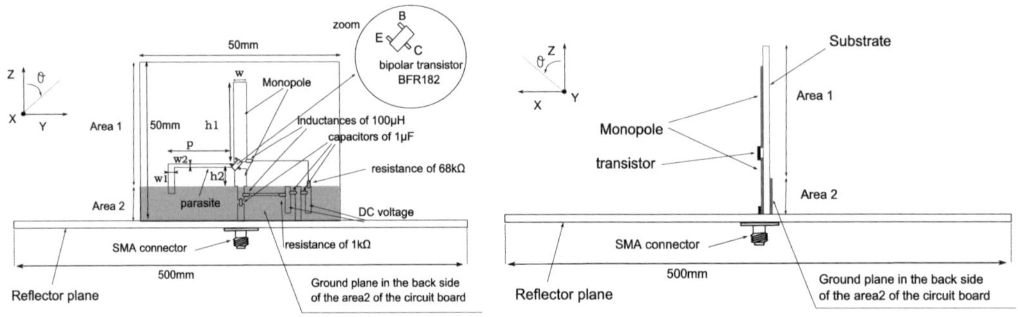

(a) Front view (b) Side view

(h_1=25mm, h_2=5mm, p=20mm, w=4mm, w_1=2mm, w_2=1mm)

Figure 2. Geometry of the proposed active receiving antenna

When V_{CC}=18V and V_{BB}=8.3V, the transistor is biased with I_C=10mA and V_{CE}=8V for common emitter configuration, I_C=4.8mA and V_{CE}=4.86V for common collector configuration. The resistances values (R_1=1kΩ, R_2=68kΩ) have been calculated in order to obtain the collector to emitter voltage (*Vce*) and collector current (*Ic)* required.

2.3. Calculation methods

We present here the theoretical and experimental methods of calculation used to evaluate the performance of the active monopole antenna. There are several methods to simulate and calculate the performance of an active antenna. Our choice is based on the method of combining the results of the electromagnetic calculation of the antenna structure with electrical model of the bipolar transistor. It conducts to a full analysis of the active antenna. This simulation was performed with CST Microwave Studio®.

With this method, only the input impedance of the active receiving antenna is calculated, because the antenna operates only at the reception due to unilaterally of the transistor. Hence, it is impossible to calculate the radiation characteristics (pattern, gain) with the usual calculate methods.

To solve this problem, we calculate the gain of the active receiving antenna by using the Link Budget method explained in [17], the setup system is illustrated in the Figure 3.

We compute the transmission parameter between a transmitting reference antenna and the active receiving antenna separated by the distance D.

Then by using the FRIIS formula (1) and knowing G_i, the transmitter antenna gain (2.1dBi), we are able to deduce G_r the gain of the active antenna.

$$S_{21} = Gi + Gr + 20\ \log\ (\lambda / 4\pi D) \tag{1}$$

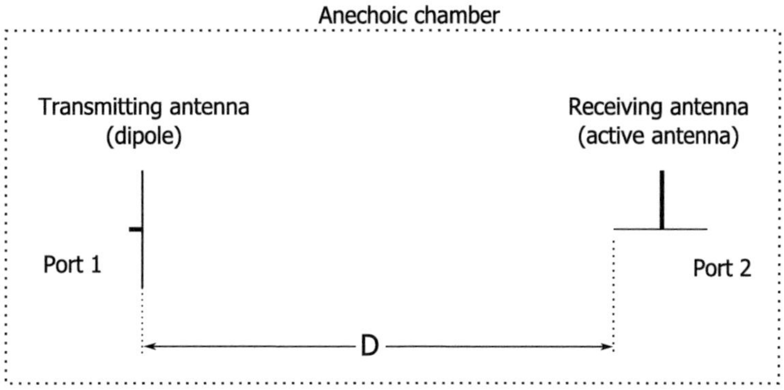

Figure 3. Two-port system proposed to calculate the active antenna performances.

Where *Gi* is the transmitter antenna gain, *Gr* the receiver antenna gain and D the distance between the two antennas. To ensure the validity of formula (1), the distance between the antennas (D) must satisfy the far field condition and the antennas are matched to 50Ω.

2.4. Theoretical results and measurements

Despite the difficulty to simulate an active antenna, we use CST software to achieve consistency between measurements and simulations. Several parameters, as the position of the transistor, the position of the parasite and the design of the antenna are investigated. All of these parameters allow us to vary the impedance of the active antenna and adjust the frequency band. In this section, we present theoretical and experimental results obtained on the active receiving antenna which operates in VHF band.

The antenna presented is working on the FM band (88-108MHz). We investigate the influence of the integration of the transistor in the monopole without matching circuit, its impact on the input impedance, and the size reduction.

We start with the influence of the height of the active monopole antenna (h_1) on its resonance frequency. Then, we investigate the influence of the transistor location on the monopole (h_2) and the position of the parasitic line (p) both on the return loss and on the gain through the calculation of the S_{21} (formula. 1) in both configurations.

2.4.1. *Influence of the height of antenna*

In this section, we present the theorical results of the performance of the active receiving antenna versus the height of the monopole (h_1), the transistor is positioned at $h_2=h_1/2$ (Figure 4). The heights of the active monopole (h_1) vary between 355 mm and 30 mm.

The theoretical results of the return loss of the active monopole as function of h_1 for both configurations, common emitter and common collector are shown in Figure 5.

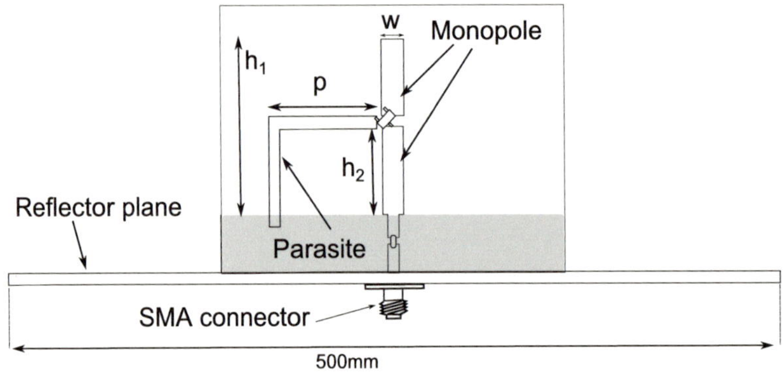

Figure 4. Geometry of the active monopole antenna.

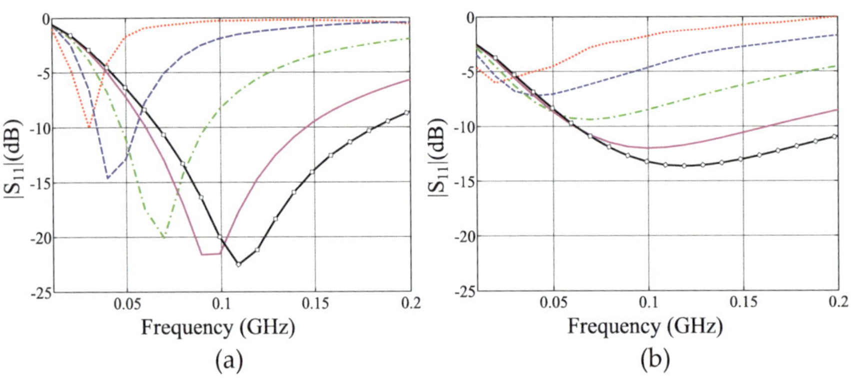

(···h_1=355mm, – – –h_1=177mm,-··-h_1=88mm, —— h_1=44mm, –▫– h_1=30mm)

Figure 5. Return loss of the active receiving antenna as function of the h_1. (a) Common emitter configuration, (b) Common collector configuration

As a result, we note a very low resonance frequency of active monopole. This resonance frequency depends on the height of the monopole and the configuration of the transistor used.

For the common-emitter configuration, the reduction factor is λ/34 for h_1=355 mm with a resonance frequency of 25 MHz and it is λ/92 for h_1=30mm with a resonance at 109 MHz. We also note that the resonance frequency of the active monopole does not vary linearly versus the length of the antenna as a liability for classical monopole. In our case, every time the height of the monopole is halved, the resonance frequency is multiplied by a coefficient which varies between 1.4 and 1.6. For the common collector configuration, we have a reduction factor of λ/42 for h_1=355mm and λ/83 for h_1=30mm.

We summarize in Table 1 the matching frequencies obtained for different heights of the active monopole for common emitter and common collector configurations.

http://dx.doi.org/10.5772/58839

h_1 (mm)	Emitter common configuration		collector common configuration	
	Fr (MHz)	Reduction ratio(λ/ h_1)	Fr (MHz)	Reduction ratio(λ/ h_1)
355	25	34	20	42
177	45	38	40	42
88	65	52.5	70	48
44	94	72.5	100	68
30	109	92	120	83

Table 1. Matching frequencies of the active monopole and size reduction ratio as function of the height h_1

This study provides the first results that we can suggest for reducing the active antennas size. In this first part, the transistor is placed at mid-height of the active monopole. To provide an accurate justification of the influence of the transistor on the miniaturization of the antenna, we will study in the next sub-section the influence of the position of the transistor on active monopole (influence of height h_2).

2.4.2. Influence of the position of the transistor

In this paragraph, we set the height of the active monopole antenna h_1 to 30 mm. We investigate the influence of the transistor location on the monopole (h_2) both on the return loss and the gain through the calculation of the S_{21}. We present the theoretical and experimental results for three different locations. The measurement process was performed in an anechoic chamber and the reference transmitting antenna was a telescopic dipole TR1722. The distance between the two antennas is 4.7m.

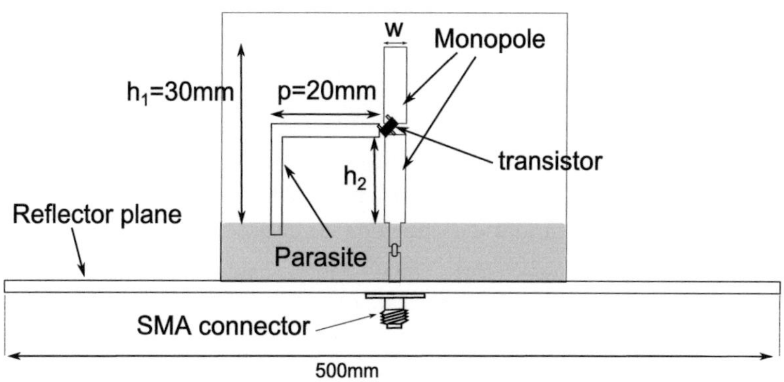

Figure 6. Position of the transistor on the active receiving antenna

2.4.2.1. Input impedance and gain

a. Common emitter configuration

The adaptation of the active antenna could be evaluated by the reflection coefficient at the output (S_{22}) of the two-port system described in Figure 3.

In Figure 7, the simulated and the measured $|S_{22}|$ are plotted for three transistor locations (h_2). Even if there is a shift between theories and measurements, the results are still in good agreement. These discrepancies are probably due to the difference between the theoretical parameters of the transistor provided by the PSPICE model and the real one. It can be noticed that the frequency bandwidth increases when the transistor is located close to the ground plane (h_2=5mm). The measurements exhibit a-10dB bandwidth of 77% around 77MHz when the transistor is located 25mm over the ground plane and 94% around 107.5MHz when the transistor is placed 5mm over the ground plane. In the first case (h_2=25mm) the height of the active antenna is close to $\lambda/212$ at the lowest operating frequency and when h_2=5mm the height of the monopole is $\lambda/175$.

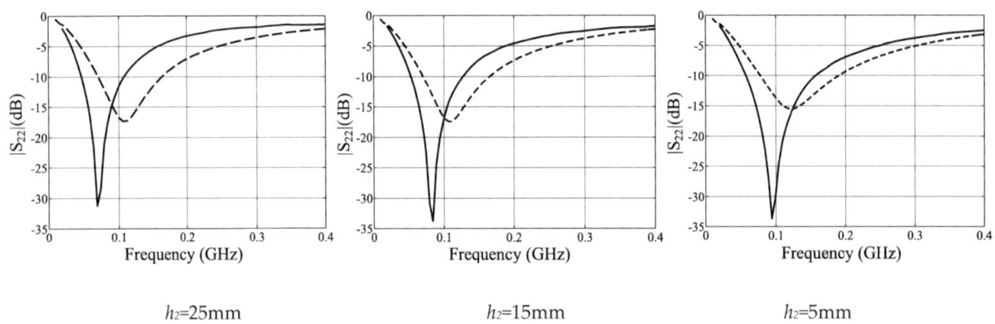

Figure 7. $|S_{22}|$ of the active receiving antenna an common emitter configuration (— measurement,---simulation)

We summarized the simulated and the measured frequency bandwidth results of the active receiving antenna in the Table 2.

For the gain of the active receiving antenna, we present the simulated and measured results of the transmission coefficient between the transmitting antenna and the active antenna for three location of the transistor (h_2=5mm, h_2=15mm and h_2=25mm). These results are plotted in Figure 8. The global shape of the $|S_{21}|$ for different transistor locations is well predicted and the theoretical results are in agreement with the measurements as a function of frequency From this results, at 130MHz, the measured transmission coefficient $|S_{21}|$ are close to-52dB and-48.17dB respectively for h_2=25mm and for h_2=5mm.

Location of the transistor	Simulated bandwidth (MHz)	Measured bandwidth (MHz)
h_2=25mm	69-162 (80%)	47-107 (78%)
h_2=15mm	70-165 (81%)	53-128 (83%)
h_2=5mm	78-190 (83%)	57-158 (94%)

Table 2. Simulated and measured bandwidth versus transistor location on the active antenna in common emitter configuration

http://dx.doi.org/10.5772/58839

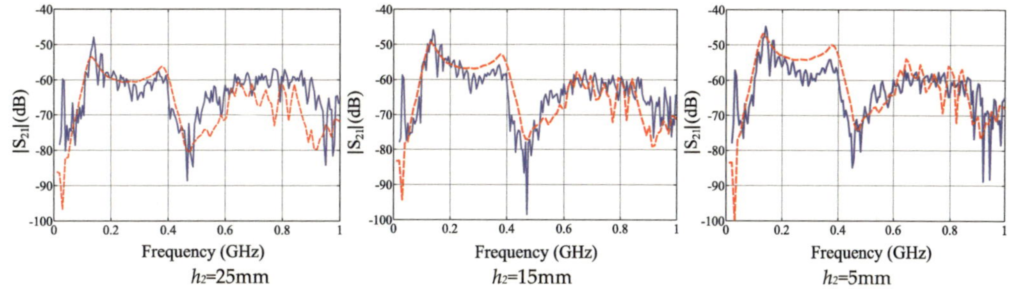

Figure 8. Transmission coefficients $|S_{21}|$ between reference antenna and active receiving antenna

From the FRIIS formula, we can estimate the measured and theoretical gain of the active receiving antenna at 130MHz. We measured a gain of-25.7dBi for h_1=25mm and-21.8dBi for h_2=5mm. The difference between these gains is closed to 3.9 dB and is due to the position of transistor on monopole. We notice that the highest gain is obtained when the transistor is very close from the ground plane (h_2=5mm). This gain level should be compared to the gain provided by an equivalent passive antenna with the same height (-49dBi).

b. Common collector configuration

In this second part, we present the same simulated and measured results of the active antenna as function of location of the transistor (h_2) for the common collector configuration. The results of $|S_{22}|$ are plotted in

Figure 9. There is a central resonance frequency around 100 MHz as in the common emitter configuration. Matching is always better in simulation and the bandwidth increases when the transistor is close to the reflector plane.

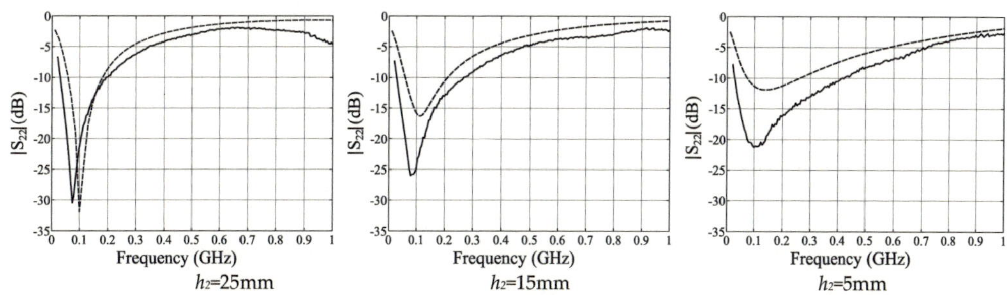

(—measurement,---simulation)

Figure 9. $|S_{22}|$ of the active receiving antenna in common collector configuration

We reported in the Table 3 the simulated and measured results.

Location of the transistor	Simulated bandwidth (MHz)	Measured bandwidth (MHz)
h_2=25mm	58-182 (103%)	29-198 (148%)
h_2=15mm	60-210 (111%)	28-272 (162%)
h_2=5mm	70-260 (115%)	27-426 (176%)

Table 3. Simulated and measured bandwidth versus transistor location on the active antenna in common collector configuration

Figure 10 presents the transmission coefficient $|S_{21}|$ as a function of the position of the transistor in common collector configuration for a height h_1=30mm. The results are presented from 50 MHz to 1 GHz frequency band to compare the behavior of the receiving active antenna as a function of frequency between simulations and measurements in order to validate the simulation methods. Agreements between the measured and simulated results are obtained. At 130 MHz, we obtained a transmission coefficient $|S_{21}|$ of-54.3 dB for a transistor positioned at h_2=25mm over to the reflector plane, of-53.8 dB for h_2=15mm and-53.8 dB when h_2=5mm.

Using the Friis formula, we calculates a gain of-28 dBi for the active receiving antenna when h_2=25mm and-27.58 dBi for h_2=5mm. The gain is almost identical between the two positions of the transistor.

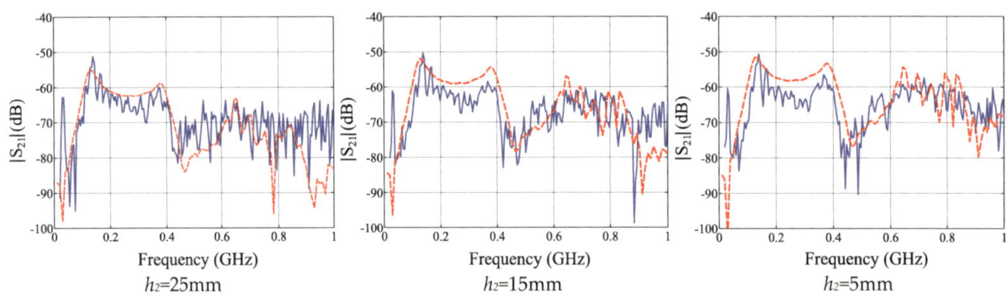

Figure 10. Transmission coefficients $|S_{21}|$ between reference antenna and active receiving antenna

These measurement results are very interesting towards the size and frequency bandwidth of the active receiving antenna and we underline that we can adjust the antenna bandwidth and the size reduction of the active receiving antenna compared to the wavelength by changing the position of the transistor.

2.4.2.2. *Radiation patterns*

To confirm our link budget calculation, we present in this part the measured radiation patterns of the active receiving antenna as function of the position of the transistor (h_2). We present the results of two heights h_2=25mm and h_2=5mm.

http://dx.doi.org/10.5772/58839

Figure 11 present the position of the active monopole over a reflector plane (500 mm x 500 mm) and the radiation pattern reference.

Figure 12 and Figure 13 show the radiation patterns of the active receiving monopole for both configurations, common emitter and common collector at several frequencies in FM band (88-108MHz). The radiation of the active antenna according to the frequency is similar to the radiation of a conventional monopole with vertical polarization.

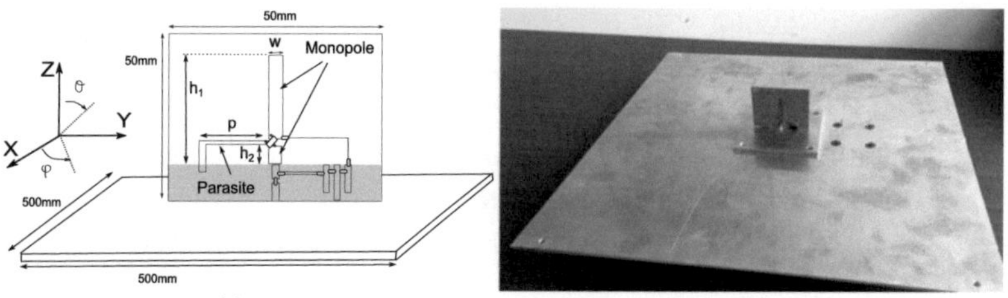

(a) (b)

Figure 11. (a) Geometry of the antenna (b) Photography of the antenna prototype on the ground plane

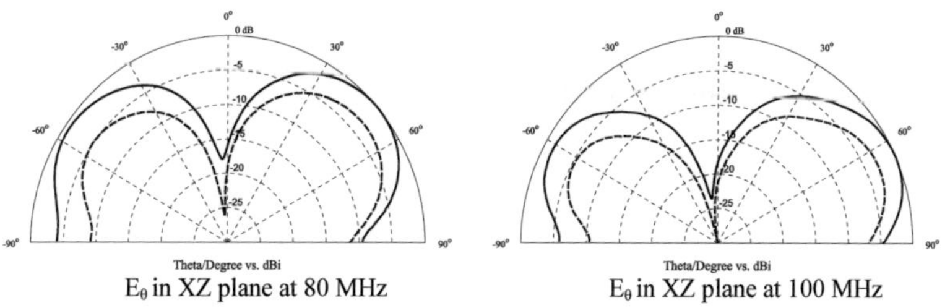

(—— h_2=5mm, h_2=25mm)

Figure 12. Normalized measured radiation patterns of the active receiving antenna in common emitter configuration

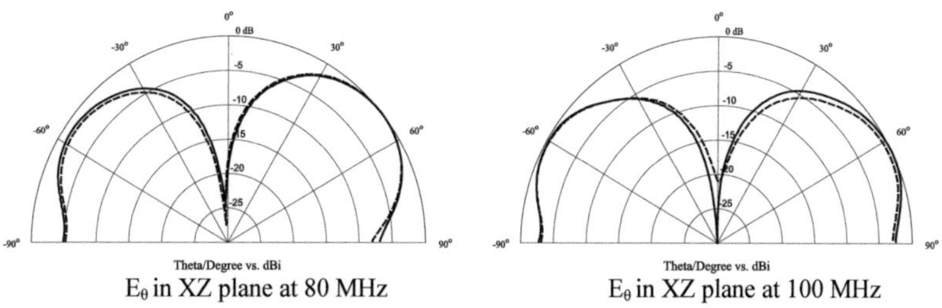

(—— h_2=5mm, h_2=25mm)

Figure 13. Normalized measured radiation patterns of the active receiving antenna in common collector configuration

As we have noticed above, the antenna gain depends on the transistor position. The gain variation between the two positions of the transistor is more important in common emitter than in common collector configuration. For example at 100 MHz, a gain of-22.6 dBi is measured for h_2=25mm and-19.6 dBi for h_2=5mm in common emitter configuration and-27.5 dBi for h_2=25mm and-27.3 dBi for h_2=5mm common collector configuration. Thus, there is a variation of 3 dB on common emitter and 0.2 dB common collector when h_2 varies from 25mm to 5mm.

The variations of transistor position provide the variation on the size reduction, bandwidth and the gain of active receiving antenna. The highest gain is obtained when the transistor is very close to the ground plane. For common emitter configuration, we have measured a gain of-19.6dBi at 100MHz and a bandwidth of 94% around 107.5MHz. In common collector configuration, the gain is equal to-27.3 dBi at 100MHz and the bandwidth is close to 176% around 226 MHz. The height of the active antenna is close to $\lambda/175$ and $\lambda/370$ in common emitter and common collector configuration respectively, where λ is the wavelength at the lowest operating frequency.

2.4.3. Influence of the position of the parasite

The study of active monopole is focused on two parameters: the height of the active receiving antenna (h_1) and the position of the transistor (h_2). In this part, the variation of the parasitic element position *p* is investigated (Figure 14). Indeed, the parasitic element is used to polarize the transistor and its length provides a variation of the input impedance of the active monopole.

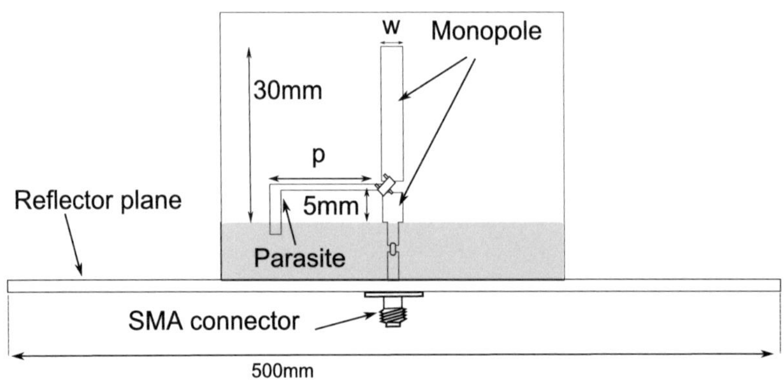

Figure 14. Position of the parasite in active antenna

2.4.3.1. Parametric studies

To observe the influence of the distance of (*p*) on the performances of the active antenna, this parameter has been sweep between 20 mm and 0.5 mm. The total length of the parasitic element varies from 25 mm to 5.5 mm. The simulation results of $|S_{11}|$ are shown in Figure 15.

http://dx.doi.org/10.5772/58839

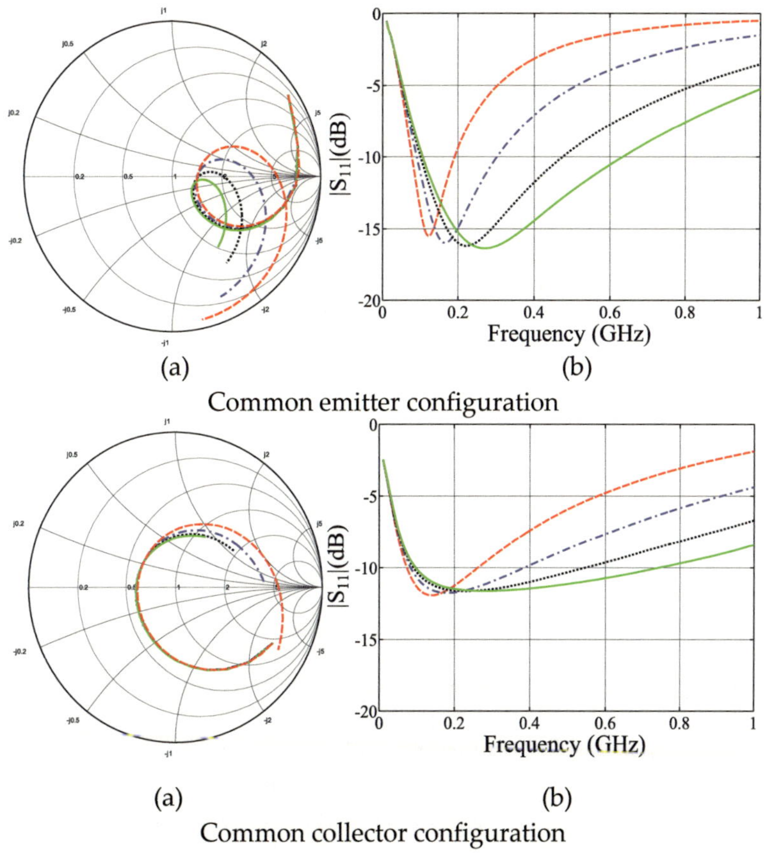

(——p=0.5mm, ··· p=5mm,-·-p=10mm,---p=20mm)

Figure 15. (a) Theoretical impedances of the active monopole as function of the position of the parasite *p* (b) Theoretical return loss active monopole based on the position of the parasite *p*.

It can be noticed that the bandwidth of the active antenna increases as *p* decreases, for the two configurations: common emitter and common collector. The results are summarized in Table 4.

Position of the parasite p (mm)	Bandwidth (MHz)	
	Common emitter	Common collector
20	78-190 (83%)	70-260 (115%)
10	90-300 (107%)	82-385 (130%)
5	100-474 (130%)	90-550 (143%)
0.5	105-632 (143%)	96-741 (154%)

Table 4. Simulated bandwidth versus parasite position (*p*) on the active antenna

2.4.3.2. Experimental results

To validate the theoretical studies of the influence of the position of the parasite on the impedance of the antenna, measurement results for two positions of the parasite in the common emitter and common collector configurations are presented. For the first one, p is equal to 20mm, for the second, p=0.5mm.

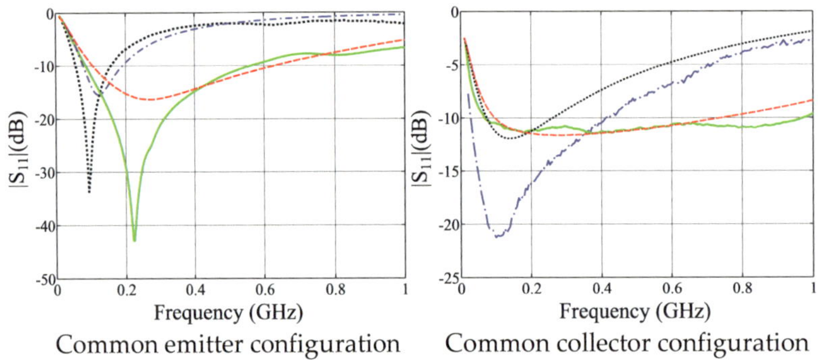

(—— measured p=0.5mm, ··· measured p=20mm,-·-·-simulated p=20mm,---simulated p=0.5mm)

Figure 16. Measured return loss of the active monopole antenna based on the position of the parasite p.

The variation of the parasitic element position can enhance the bandwidth of active receiving antenna as shown in Figure 16. For the common emitter configuration, we measured a bandwidth of 94% around 107 MHz for p=20 mm and a bandwidth of 149.5% around 320 MHz for p=0.5mm. For the common collector configuration, when p=0.5mm, there is a-10 dB bandwidth of 175% around 521 MHz measurement (Table 5). For the two configurations, the measurements and simulations are in good agreements.

	Emitter common configuration		Collector common configuration	
	Simulated	Measured	Simulated	Measured
p=20mm	78-190	57-158	70-260	27-426
p=0.5mm	105-636	81-560	96-741	65-978

Table 5. Simulated and measured bandwidth (MHz) versus parasite position (p) on the active antenna

2.4.4. Influence of the geometry of the monopole

In the first part of this chapter, we have presented the influence of different parameters on the performance of an active receiving monopole. The influence of the transistor position and the parasitic element position on the size reduction, on the bandwidth and on the gain has been clearly underlined.

To improve the gain performance of the active monopole antenna, we change the geometry of the antenna, especially on the upper part of the monopole. The transistor is in common emitter configuration positioned at h_2=5mm (h_1=30mm, p=0.5mm).

We calculated and measured two structures of active monopole antenna (Figure 17), the length of the upper part of the monopole ranges from 25 mm to 106 mm.

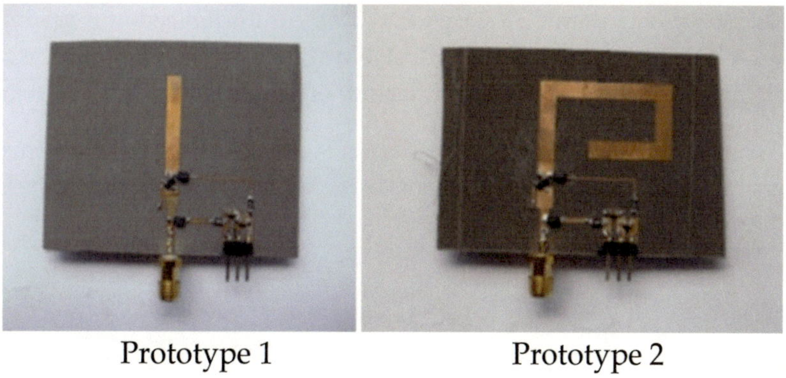

Figure 17. Geometry of the active monopole antenna

The theoretical and experimental return losses of the active antenna as a function of the length of the upper part of the monopole are presented in Figure 18.

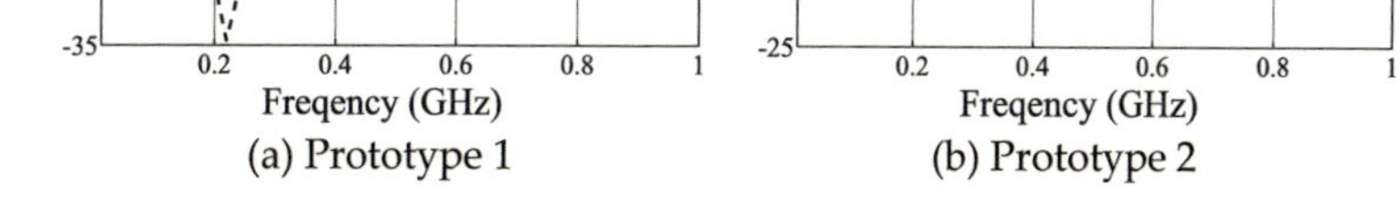

(---measured, —— simulated)

Figure 18. return losses of active monopole antenna

There is a frequency shift between measurements and simulations. However, measurements and simulations are in a good agreement. We measured a bandwidth of 150% around 322 MHz for prototype 1 and 130 % around 228 MHz for prototype 2. These results are summarized in Table 6.

	simulated BW (MHz)	measured BW (MHz)
Prototype 1	108-693 (146%)	81-563 (150%)
Prototype 2	97-335(110%)	80-377(130%)

Table 6. Simulated and measured bandwidth of the active antenna

In Figure 19, we presented the measured radiation patterns of the proposed antenna at 80 MHz and 100MHz. Radiation patterns are in good agreement for the bending structure and the classical one. A maximum gain of-21.57 dBi at 100 MHz was measured for the prototype 1 and-15.11 dBi for prototype 2. We have a difference of 6 dB between the two prototypes.

For evaluate the performances of active antenna in FM radio reception, we have measured the received signal strength indicator and the signal to noise ratio in FM band using the FM receiver evaluation board of silicon labs (Si4706).

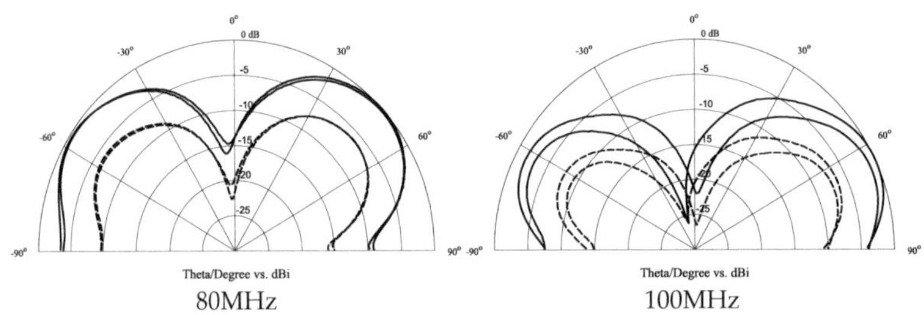

(---prototype 1, —— prototype 2)

Figure 19. Measured normalized radiation patterns of the active antenna in the XZ and YZ plan

The results are plotted for the active receiving antenna (prototype 2) and a reference monopole antenna of 60cm of height. The average power received by the reference $\lambda/4$ monopole antenna is 10dB higher than the power received by active receiving antenna. Despite the reduction of the power level, we measured a good received signal quality for listen to FM radio without interference.

2.5. Conclusion

In this first section, we studied the influence of the integration of a transistor on a passive monopole antenna, including the miniaturization of the antenna and the increase of the frequency bandwidth. Our contribution is based on the results obtained by Meinke in 60-70th and we conducted a theoretical and experimental validation of the active receiving antennas.

Despite the difficulty to simulate an active antenna, we use CST software to achieve consistency between measurements and simulations. Several parameters, as the position of the transistor, the position of the parasite and the design of the antenna are investigated.

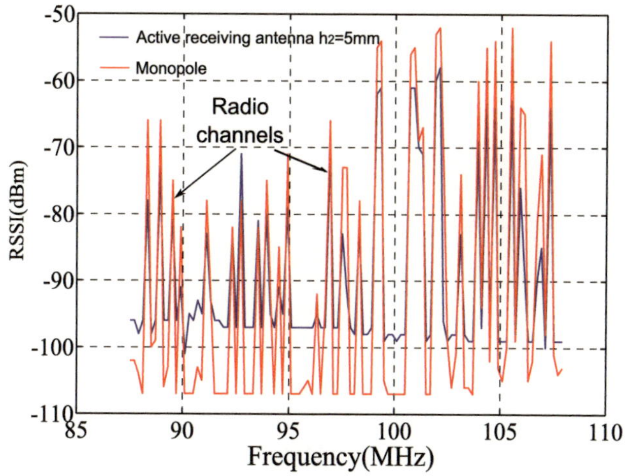

Figure 20. Received Signal Strength Indicator in FM band.

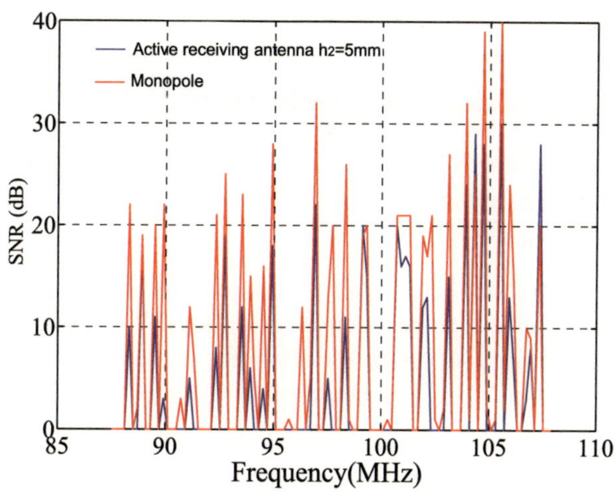

Figure 21. Signal to Noise Ratio

We also studied two configurations of transistor, i.e. common emitter and common collector. It was found that each configuration provides different performances. The bandwidth is wider in common collector (176%) than common emitter (94%). The reduction size is equal to λ/370 in common collector configuration and λ/175 in a common emitter.

Concerning the gain, the common emitter configuration presents a gain of-19.6 dBi which is higher than the gain (-27.3 dBi) obtained with the common collector configuration. These gains must be compared to gain a classical monopole of the same height (30mm) which is-49dBi.

Finally, the change of active monopole geometries has allowed us to increase the gain of the active antenna. It led to the creation of a miniature active antenna with a bandwidth of 130% around 228 MHz, the height of the active antenna is $\lambda/80$ at the lowest frequency of the bandwidth. The measured gain is-15.1 dBi at 100 MHz. We have measured a good received signal quality.

3. Compact tunable antenna

3.1. Introduction

Over the past years, efforts have been made to design small antennas in the UHF band for handset applications. Because of the convergent trend, limited space is available for each antenna device. For applications like broadcast reception, the antenna should be able to cover 40% of the relative bandwidth from 470 to 702 MHz. Regarding the wavelength at the lowest part of the bandwidth ($\lambda_{0=}$638 mm), it is obvious that small antennas are required for such applications.

To meet this requirement, the strategy is to cover the frequency range of such broadband applications using frequency tunable antennas. The basic antenna is then narrow band and the implementation of active devices introduces a frequency tunable ability. From a system point of view, it provides frequency selectable functions, which improve the signal-to-noise ratio. In [20-22], the authors show a compact tuned antenna for mobile applications. In [20,21], varactor diodes have been associated with PIFAs or meander antennas. Radiation patterns and S_{11} demonstrate the antenna's performances. In [23], an RF switch has been chosen rather than a varactor diode to avoid radiation influence on the device. The received measurement results show the validity of the concept.

In this section, we will consider the theoretical and physical aspects of self-inductance and investigate its effect on the resonance frequency of the MCLA (Monopole Coupled Loop Antenna) [24]. We studied them in order to explain the evolution of size reduction and the radiation pattern. We obtained a 55% reduction in the size of the antenna compared to the initial MCLA.

Finally, we associate a varactor diode with a small modified open-circuit MCLA in order to continuously control the frequency with a DC bias voltage. As a result a desirable frequency in a broadband frequency range covering 470–675 MHz is achieved. Experimental and theoretical results including S_{11}, radiation patterns and gain are in good agreement.

3.2. Monopole Coupled Loop Antenna (MCLA)

In this paragraph, we propose to feed a short circuited printed half loop antenna through an electromagnetic coupling using an arc monopole line [24]. The half loop, short circuited on its both sides, has been associated to an arc monopole fed by a 50Ω SMA connector as described in Figure 22. The antenna has been printed on a dielectric substrate circuit board and mounted above a finite reflector ground plane. The theoretical antenna performances have been computed with CST Microwave Studio Software and compared to measurements. Regarding this first part of the study, we propose a physical explanation of the antenna behavior. Thereafter, parametric studies give additional information and help us to increase our knowledge of this design.

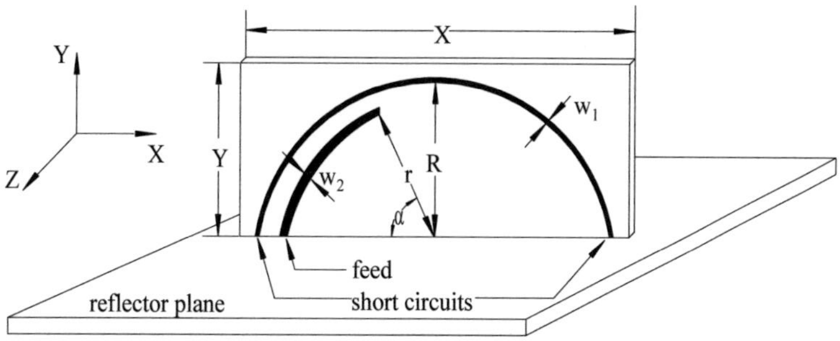

Figure 22. Geometry of the proposed antenna

R and r are respectively the radii of the half loop and the arc monopole lines, w_1 and w_2 are their widths. P is the total length of the arc monopole line and α is the angle between its both extremities. The radius (thus the length) of the half loop line will remain constant.

By an optimized choice of the lengths, the widths and the distance between the two lines, it is possible to achieve a broadband solution presented in Figure 23. The antenna has been printed on a Neltec NY9300 substrate (ε_r=3, h=0.786 mm, tanδ=0.0023) and above a limited square ground plane (300×300×4 mm^3).

In this case, R=100 mm, r=84 mm, w_1=0.4 mm, w_2=0.5 mm, X=220 mm, Y=110 mm, α=90° and P=132 mm. The return loss and the input impedance of the antenna have been computed. The simulation and measurement have been performed between 200MHz and 800MHz.

The Figure 23 shows a comparison between simulation and measurement with a very good agreement. The measured return loss bandwidth is close to 70MHz (≈15.3%). With this design, we increase three times the bandwidth of the antenna proposed in [25].

On Figure 23.a, the shape of the return loss shows two resonances. The first resonance frequency depends of the $\lambda/2$ half loop radiator, whereas the second one has been created by the arc monopole considered as a quarter wavelength conventional monopole. As noticed in reference [26], the electromagnetic coupling between the two parts of the antenna affects the impedance behavior.

To increase our understanding of this antenna, we proceed to theoretical parametric studies. We investigate the dimensions of the ground plane, the length, the width of the arc monopole line, the width of the half-loop line and the distance between the two microstrip lines and we will show the influence of these parameters on the input impedance. In this section, we present just two parameters, the length of the arc monopole and the distance between the two printed lines.

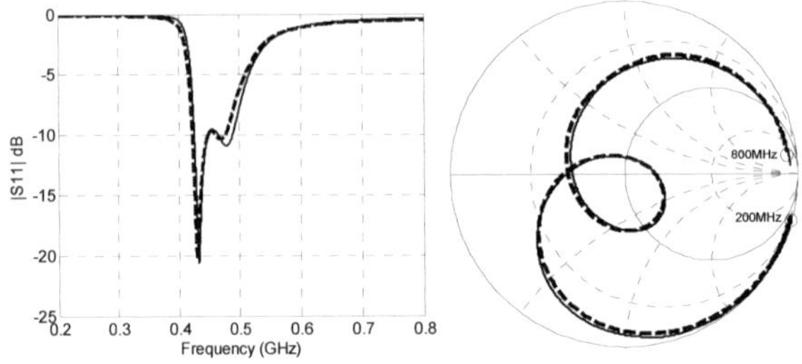

(a) (b)

(---) measured, (—) simulated

Figure 23. (a) Simulated and measured return losses for the MCLA, (b) Simulated and measured antenna input impedances

3.2.1. Length of the arc monopole

In Figure 24, we present the input impedance of the MCLA versus α which is the angle between the extremities of the arc monopole. α varies from 64.8° to 133°. We notice that there are no modifications on the other half loop parameters. The circles on the curves represent the resonance frequencies of the arc monopole (Figure 24).

The resonance of the half loop remains constant (Figure 23). The resonance mode of the arc monopole decreases when the length of this line increases. The best case is for α=90°, the resonance frequencies of the two lines are quite close and the impedance matching criterion S_{11}<-10 dB has been used to calculate the impedance bandwidth (70MHz or 15.3%). In the other case, the lower resonance frequency is too far from the highest one and it introduces a mismatching phenomenon. In Figure 24(a), for α=103.7°, both resonance frequencies are so close that they seem to be a single one. On Figure 24(b), we can distinguish these two resonances.

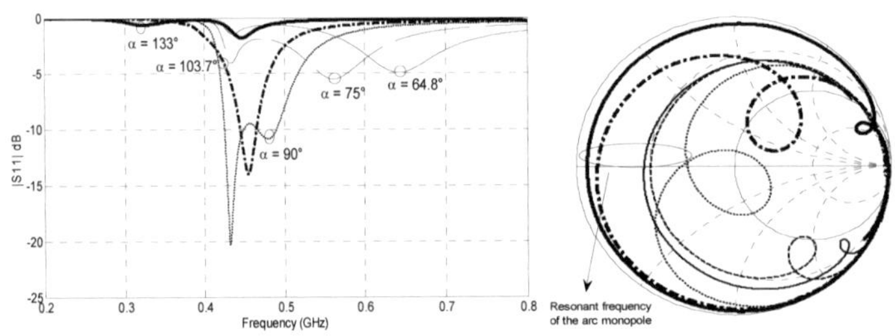

(a) (b)

(▬) α=133°, (- -) α=103.7°, (···) α=90°,(----) α=75°, (—) α=64.8°.

Figure 24. (a) Return losses of the simulated MCLA versus α (b) simulated MCLA input impedances, the frequency is varying between 200 and 800MHz.

3.2.2. Distance between the two printed lines

In Figure 25, three values of the arc monopole radius *r* have been used to characterize the MCLA. *r* varies from 70 mm to 95 mm and α is variable to keep the same length of the arc monopole. *R*=100 mm, w_1=0.4 mm, w_2=0.5 mm, *P*=132 mm, *X*=220 mm and *Y*=110 mm.

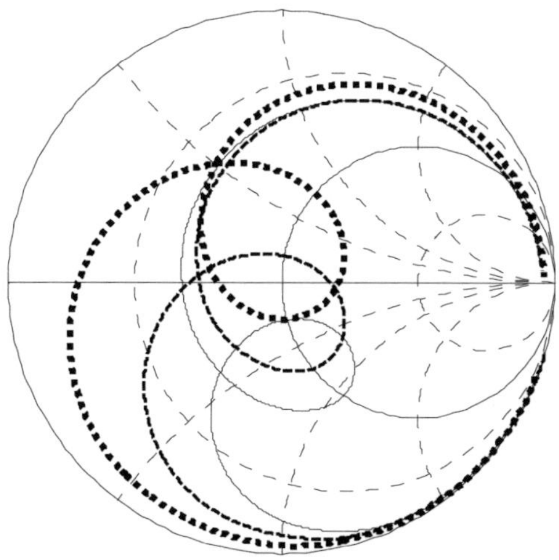

(—) *r*=70mm, (----) *r*=84mm, (···) *r*=95mm

Figure 25. Simulated MCLA input impedances when the radius of the arc loop is varying between 70mm and 95mm (the other parameters are constant), the frequency is varying between 200 and 800MHz.

The resonance frequency of the arc monopole decreases when the radius of this line increases. The best case is for *r*=84mm. These variations demonstrate that the coupling effect is very sensitive to the distance between the half loop and the arc monopole. The resonance of the half loop remains constant.

A monopole coupled loop antenna (MCLA) has been proposed [24]. A better impedance bandwidth has been obtained by electromagnetic coupling effect between the two microstrip lines (arc monopole+half loop). Its bandwidth is three times wider than the conventional half loop short circuited monopole [25]. In the next paragraph, different techniques will be used to reduce the size of the MCLA.

3.3. MCLA loaded by self-inductance

To achieve our first purpose (size reduction of the antenna), we modify the MCLA presented in previous paragraph by loading the short circuit of the half-loop on the left side with a self inductance (Figure 26). The ground plane is an infinite one [27].

The main dimensions of the antenna are listed below: R=100mm, r=84mm, w_1=0.4mm, w_2=0.5mm, Y=220mm, Z=110mm, α=90°.

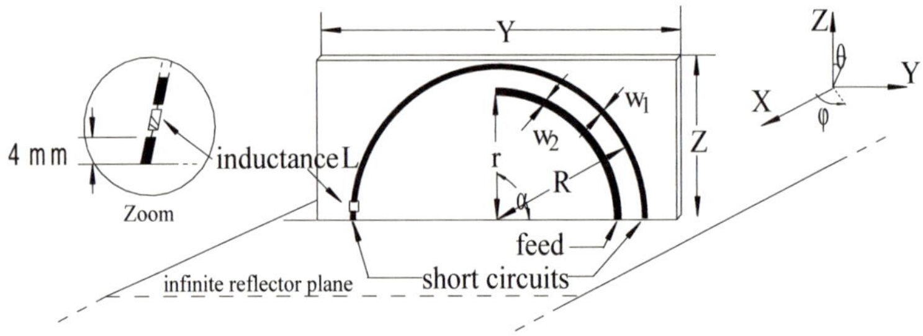

Figure 26. Geometry of the proposed antenna

The inductor component used below is an ATC 0805 WL modelled with a parallel RLC device. Simulations have been performed with CST Microwave studio® where the equivalent RLC circuit model has been introduced as lumped components [27].

In Figure 27, we present the input impedance of the MCLA loaded versus self inductance value. The self values vary from 0nH (equivalent to short circuit) to an infinite value (equivalent to an open circuit).

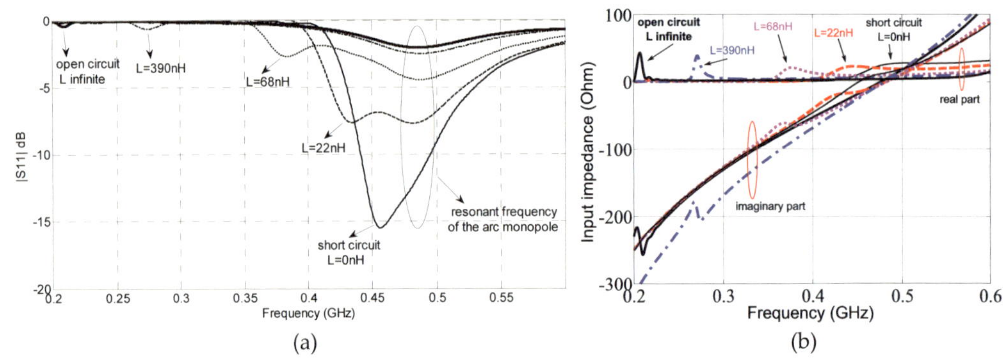

Figure 27. (a) Simulated return losses for the proposed antenna with different values of the self inductance, (b) Simulated input impedance for the proposed antenna with different values of the self inductance

In Figure 27, for each inductance, we can notice two resonance frequencies. The first one, linked to the arc monopole, remains constant and very close to 450MHz, whatever the self inductor values. The second one linked to the half loop radiator associated to the self inductor decreases from 455MHz to 209MHz. We can also notice that for these resonances, the antenna becomes more and more mismatched when the inductance value increases. In the previous paragraph we have mentioned that the antenna return losses at these resonance frequencies could be matched by modifying the length of the arc monopole.

To complete this calculation, we have computed and represented the normalized radiation patterns of the short circuited MCLA (L=0nH) (Figure 28).

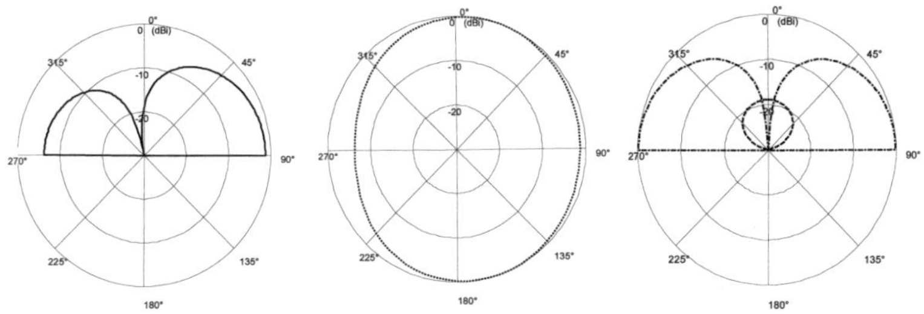

(a) (b) (c)

Figure 28. Normalized radiation patterns of the short circuited MCLA (L=0nH). (a) E_θ in YZ-plane φ=90°, (b) E_θ in XY-plane θ=90°, (c) (----) E_θ in XZ-plane and (- -) E_φ in XZ-plane φ=0°

For YZ-plane (Figure 28.a), we obtain a dissymmetric shape of radiation pattern due to the arc monopole influence [24]. In XY-plane, the radiation pattern is quasi-omnidirectional. Except the dissymmetry of the radiation pattern in YZ plane, we can notice that the general behavior of the antenna is very close to this provides by a monopole. The gain is equal to 1.8 dBi at 460 MHz.

This analysis is completed by drawing surface currents on the radiating elements (Figure 29).

Near to the feed point and to the short circuit areas, the surface currents are maximum. It explains that the maximum radiations are obtained for the E_θ in YZ-plane when θ=90°, and for the E_θ in XZ-plane when θ=90° or 270°. In Figure 29, the surface currents become null in z direction.

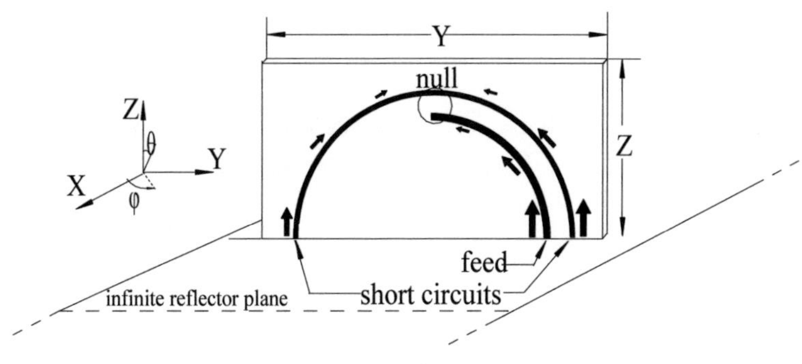

Figure 29. Schematic of surface currents distribution on radiator elements of short circuited MCLA

By increasing the value of the inductor, the deep null in the radiation pattern in YZ plane disappears and the antenna provides a quasi unidirectional pattern in this plane. This phenomenon could be explained in the Figure 30 where surface currents have been represented when the inductance value is 390 nH. Then, we notice that the half-loop surface currents distribution shifts to the left.

The inductance modifies the equivalent antenna electrical length and we can deduce that the antenna operating wavelength is lower than half a wavelength. In the extreme case, where the self inductance is infinite and equivalent to an open circuit, the open circuited MCLA operating wavelength becomes equal to a quarter wavelength.

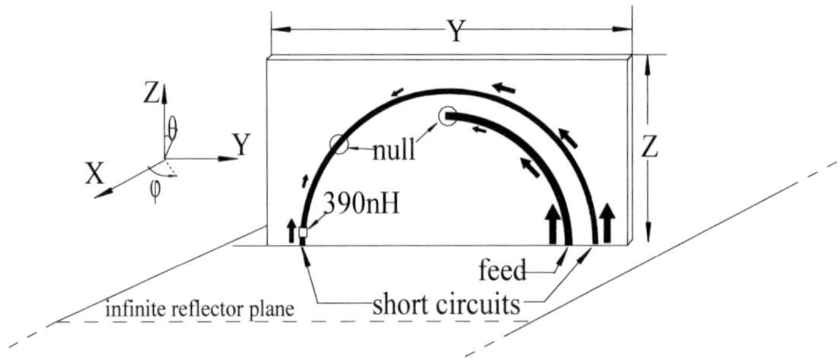

Figure 30. Schematic of surface currents distribution on elements of the MCLA loaded by a 390nH inductance

3.4. Modified open-circuit MCLA

We keep the case of the modified MCLA with an infinite self inductance [27]. We call this antenna modified open circuit MCLA. As mentioned in the previous paragraph and in [24], we could modify the arc monopole length to enhance the impedance bandwidth. As we wish that the antenna operates at 450MHz, we have optimized the arc monopole length to achieve both impedance matching and size reduction.

In Figure 31, we remind the modified open circuited MCLA design. For practical reasons, the antenna is placed perpendicularly above a limited square ground plane (300mm x 300mm x 4mm).

The main dimensions of the antenna are listed below: R=44.5mm, r=40.5mm, w_1=1mm, w_2=2mm, Y=100mm, Z=50mm, α=178.5°. The reduction size of the modified antenna is 55.5% compared to the initial MCLA.

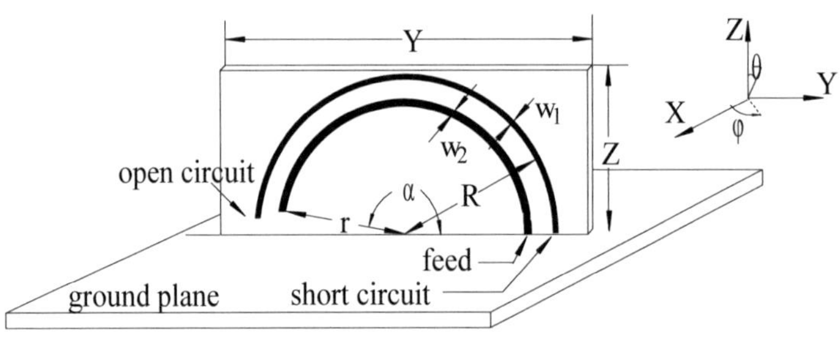

Figure 31. Geometry of the modified open circuited MCLA

http://dx.doi.org/10.5772/58839

The Figure 32 shows the theoretical and the measured return losses and input impedance with a good agreement.

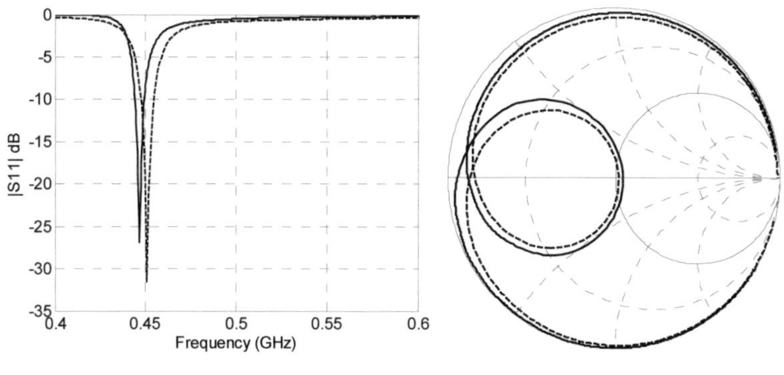

(a) (b)

(---) measured, (—) simulated

Figure 32. (a) Simulated and measured return losses for the modified open circuit MCLA (b) Simulated and measured antenna input impedances.

In Figure 33, we present the measured and simulated normalized radiation patterns of the modified open circuit MCLA.

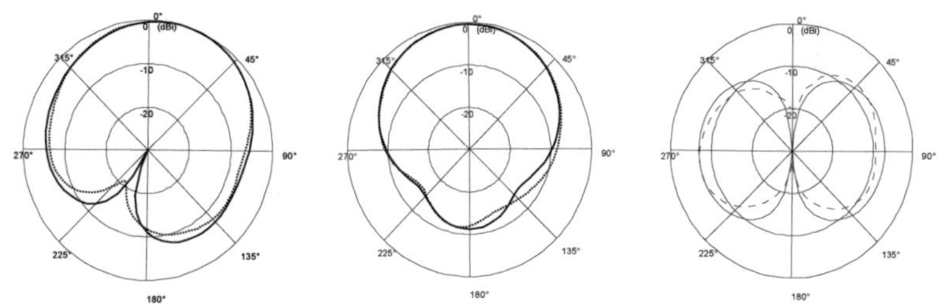

E_θ YZ plane E_φ XZ plane E_θ in XZ-plane

(---) measured, (—) simulated

Figure 33. Normalized radiation patterns at 448 MHz

As expected, the radiation pattern is quasi unidirectional and the agreement between theory and experience is good. At 448 MHz, the measured and the theoretical gain are respectively 3.8dBi and 3.9 dBi and the comparison is also good.

While the inductor element is well-known to decrease the antenna resonance frequency, based on MCLA, a theoretical study has been performed to demonstrate the modification of radiating behavior of MCLA loaded with inductance. Thus a modified open circuit MCLA has been proposed. This antenna provides quasi unidirectional radiation pattern and size reduction (55.5%) compare to the initial MCLA [24]. The measured gain is high (3.8dBi).

3.5. Frequency tunable MCLA

In this section we present a frequency tunable antenna made with an open circuit Monopole Coupled Loop Antenna (MCLA) associated to a varactor diode [29]. The proposed antenna shows a 35,8% relative bandwidth, covering the [470-675]MHz frequency range.

We investigate the capability to obtain a frequency tunable antenna with reduced size open circuit MCLA. The antenna resonance frequency is controlled with a varactor diode fed having a DC bias voltage range of [0-5]V. This low voltage requirement made this diode suitable for handset application.

Performances of tunable antenna have been first investigated using CST® software, substituting the diode by its equivalent circuit. The varactor component used below is a MA4ST2200 from MACOM. The equivalent circuit proposed by the vendor is a series RLC circuit with a parallel capacitance (Figure 34). C_s is the tunable capacitor. The varactor has been characterized using a vector network analyzer and then the equivalent circuit has been obtained through a de-embedding process. In order to keep an operating frequency close to 470 MHz when the DC bias voltage is equal to zero, the parameters of the modified MCLA open circuit have been optimized with the corresponding diode equivalent circuit characteristics.

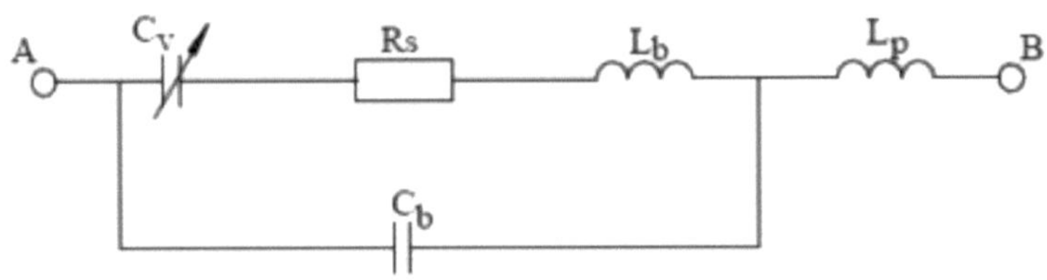

Figure 34. Equivalent circuit model of the varactor diode

A sketch of the prototype is reported in Figure 35, including the diode located on the short circuit side of the MCLA (right hand side of the drawing). The antenna has been printed on a Neltec NY9300 circuit board ($\varepsilon_{r=}3$, h=0.786 mm) and is placed perpendicularly to a square ground plane (300 mm x 300 mm x 4 mm). R and r are the radius of the half loop and the arc monopole lines respectively, w_1 and w_2 are their widths. α is the angle between the arc monopole extremities. The main dimensions of the antenna are listed below: R=37.5 mm, r=34.5 mm, w_1=1 mm, w_2=2 mm, Y=80 mm, Z=40 mm, α=43.5°.

The bias circuit used to carry a DC control voltage to the varactor without interfering with the high frequency currents is reported in Figure 35. In order to isolate the DC voltage from RF signal a choke L=2200 nH, a series resistance 5 kΩ and one chip capacitor (5 nF) are used.

http://dx.doi.org/10.5772/58839

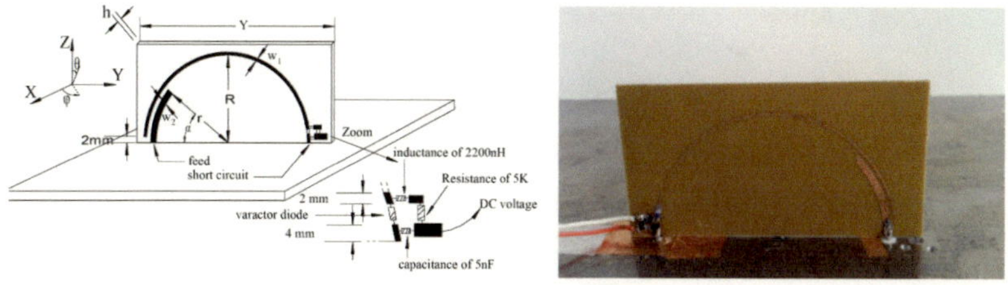

(a) (b)

Figure 35. (a) Geometry of the open circuit MCLA (b) Photography of the antenna prototype

Antenna performances for different DC control voltage values obtained with CST software have been compared to measurement results. In Figure 36, good agreement between simulated and measured return loss for a few voltage values can be seen.

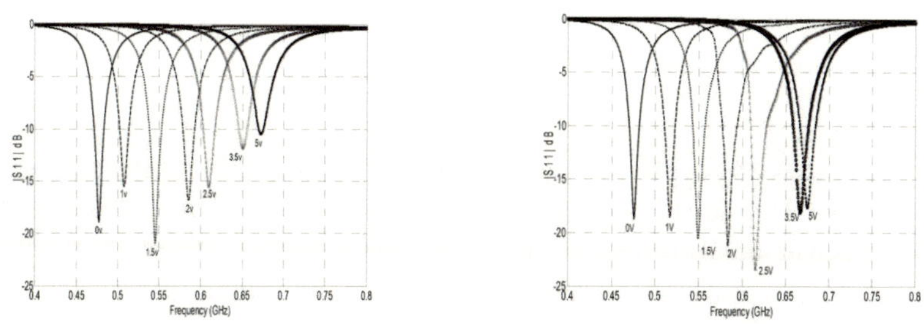

(a) (b)

Figure 36. (a) Simulated and (b) measured return losses

As depicted in Figure 36, the antenna covers the [470-675] MHz frequency range, which corresponds to a 35,8% relative-10dB adaptation bandwidth. For each resonance, the corresponding instantaneous bandwidth is more than 10MHz.

Normalized radiation patterns comparison for the frequency range extrema values 474MHz and 668MHz, corresponding to 0V and 5V DC order, show as well acceptable agreement. It has to be noticed that the behavior of the antenna radiation patterns in the main planes (YZ and XZ planes) is quasi unidirectional in the entire band. The cross-polarization in the XZ-plane becomes very low when the resonance frequency increases. The front to back radiation level decreases when the resonance frequency increases.

Good agreement is obtained for simulated and measured maximum gains versus operating frequencies (Figure 37). It has to be underlined that considering the antenna electrical dimensions ($\lambda/17 \times \lambda/8$ at 470MHz), a high gain on the whole frequency range is provided, and especially in the higher band.

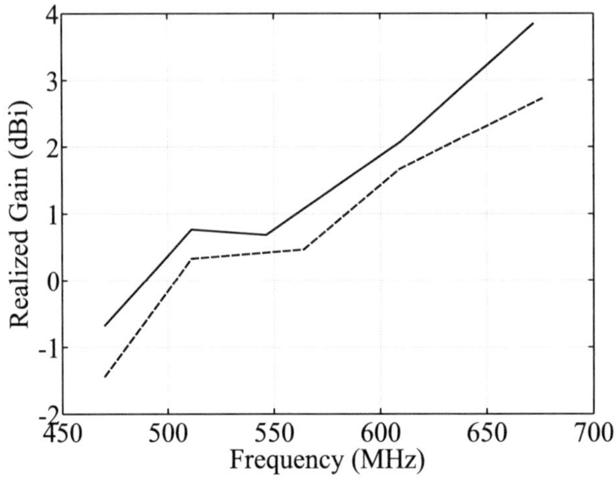

(---) measured (—) simulated

Figure 37. Simulated and measured maximum gains

3.6. Conclusion

A varactor-tuned open circuit MCLA has been designed and realized. The aim of this study is to provide a simple technique to sweep the instantaneous bandwidth of the structure from 470 MHz to 675 MHz. The measured maximum gain varies between-1.5dBi to 2.8dBi. In future, this antenna can be miniaturized in order to be included in a terminal mobile. Theoretical and experimental results of S_{11}, radiation patterns and gain have been performed and show good agreement.

3.7. General conclusion

The objectives of this chapter concern the design and the development of compact active antennas, working on a wide frequency band. We have presented innovative techniques to reduce the antenna size with a wide cover frequency rang. We have demonstrated two solutions; the first one is broadband antenna with very important size reduction and wide bandwidth. The second one is a tunable antenna.

In the first part, the capabilities offered by active antennas for miniaturization by the integration of actives components directly on the antenna have been presented and their impact on printed antenna's performances (bandwidth, size reduction, and gain) investigated theoretically and experimentally. Regarding the small volume dedicated for antenna in devices, a broadband miniature active antenna operating in the FM band has been presented.

In the second section, a technique of narrowband antennas miniaturization for DVB-H applications has been proposed. Some new designs of miniaturized antennas based on a coupling feeding system have been proposed and fabricated. Thus, these antennas have been associated with varactor diodes in order to achieve frequency tunable antenna.

http://dx.doi.org/10.5772/58839

Author details

Y. Taachouche, M. Abdallah, F. Colombel, G. Le Ray and M. Himdi*

*Address all correspondence to: mohamed.himdi@univ-rennes1.fr

Institute of Electronic and Telecommunication of Rennes (IETR), University of Rennes , Rennes, France

References

[1] H. A. Wheeler, 'Fundamental limitations of small antennas', Proc. IRE, 35, pp. 1479-1484, Dec. 1947.

[2] H. A. Wheeler, "Small Antennas" IEEE Trans. Antennas Propagat., vol. 23, pp. 462-469, July 1975.

[3] Jenshan. Lin, Itoh, T, "Active Integrated Antennas", Microwave Theory and Techniques, IEEE Transactions on, vol. 42, issue 12, pp. 2186-2194, Dec 1994.

[4] H.H. Meinke. "Transistorized receiving antennas" Institut fur Hochfrequenztechnik der Technischen Hochschule, Munchen, November 1967, 99 pages.

[5] J. R. Copeland, W. J. Robertson, R. J. Verstraete, "Antennafier arrays", Antennas and Propagation, IEEE Transactions on, vol 12, Issue 2, pp. 227-233, Mar 1964.

[6] F. M. Landstorfer, H. H. Meinke, "Transistorized Microwave Antenna with 1GHz Centre frequency", Microwave Conference, 2nd European, vol 1, pp.1-4, 1971.

[7] Anderson. A, Davies. W, Dawoud. M, Galanakis. D, " Note on Transistor-Fed Active-Array Antennas", Antennas and Propagation, IEEE Transactions on, vol 19, Issue 4, pp. 537-539, 1971.

[8] Ramsdale. P.A, MacLean. T. S. M, "Active Loop-Dipole Aerials", Electrical Engineers, Proceedings of the Institution of, vol 119, issue 4, pp 423-424, 1972.

[9] Rangole. P.K, Saini. S.P.S, "Transistor Configurations in Integrated Transistor Antennas", Radio and Electronic Engineer, vol 45, issue 3, 1975.

[10] Ramsdale. P.A, MacLean. T.S.M, "Active Loop-Dipole Aerials", Electrical Engineers, Proceedings of the Institution of, vol 118, issue 12, pp. 1698-1710, 1971.

[11] Rangole. P.K, Midha. S.S, "Short antenna with active inductance", Electronics Letters, vol 10, issue 22, pp. 462-463, 1974.

[12] Y. Qian, Tatsuo Itoh, "Progress in active integrated antennas and their applications" IEEE transactions on microwave theory and technique, Vol. 46, No. 11, Nov 1998, pp. 1891-1900.

[13] T.S.M MacLean and G. Marris, "Short rang active transmitting antenna with very large height reduction". IEEE. Transactions on antennas and propagation. March 1975, pp. 286-287.

[14] A.P.Anderson, M. Dawoud, "The performance of transistor fed monopoles in active antennas" IEEE transactions on antennas and propagation, Vol:21, issue:3, may 1973, pp. 371-374.

[15] V. B. Ertürk, R. G. Rojas, and P. Roblin, "Hybrid analysis/design method for active integrated antennas," IEE Proc.-Microw. Antennas Propagat., vol. 146, pp. 131-137,1999.

[16] P.S. Hall, "Analysis of radiation from active microstrip antennas" Electronics letters, 7th january 1993, vol 29, n°1, pp. 127-129.

[17] H. An, B. K. J. C. Nauwelaers, A. R. Van de Capelle, R. G. Bosisio, "A novel measurement technique for amplifier-type active antennas". Microwave symposium digest, IEEE MTT-S international. 1994, Vol3, pp. 1473-1476.

[18] Taachouche, Y.; Colombel, F.; Himdi, M., "Influence of the transistor location on the behavior of a transistorized printed antenna," *Antennas and Propagation (EUCAP), 2012 6th European Conference on,* vol., no., pp.1255,1258, 26-30 March 2012

[19] Y. Taachouche, F. Colombel, and M. Himdi, "Very Compact and Broadband Active Antenna for VHF Band Applications," International Journal of Antennas and Propagation, vol. 2012, Article ID 193716, 4 pages, 2012

[20] Nguyen, V.-A., Dao, M.-T., Lim, Y.T., and Park, S.-O.: 'A compact tunable internal antenna for personal communication handsets', IEEE Antennas Wire. Propag. Lett., 2008, 7, pp. 569–572

[21] Komulainen, M., Berg, M., Jantunen, H., and Salonen, E.: 'Compact varactor-tuned meander line monopole antenna for DVB-H signal reception', Electron. Lett., 2007, 43, (24), pp. 1324–1326

[22] Yoon, I.-J., Park, S.-H., and Kim, Y.-E.: 'Frequency tunable antenna for mobile TV signal reception'. IEEE AP-S Int. Symp. Dig., 2007, pp. 5861–5864

[23] Suzuki, H., Ohba, I., and Minemura, T.: 'Frequency tunable antennas for mobile phone for terrestrial digital TV broadcasting reception'. IEEE AP-S Int. Symp. Dig., Singapore, 2006, pp. 2329–2332.

[24] Abdallah, M.; Colombel, F.; Le Ray, G.; Himdi, M., "Novel Printed Monopole Coupled Loop Antenna," *Antennas and Wireless Propagation Letters, IEEE,* vol.7, no., pp. 221,224, 2008.

[25] H. Lebbar, "Analyse et conception d'antennes imprimées multifilaires", phd thesis, University of Rennes 1, France, october 1994.

[26] Z. N. Chen and Y.W.M. Chia, 'Broadband monopole antenna with parasitic planar element', Microwave Opt. Technol. Lett., Vol.27, No. 3, pp. 209-210, Nov. 2000.

[27] Abdallah, M.; Colombel, F.; Le Ray, G.; Himdi, M., "Quasi-Unidirectional Radiation Pattern of Monopole Coupled Loop Antenna," *Antennas and Wireless Propagation Letters, IEEE*, vol.8, no., pp.732,735, 2009.

[28] Taachouche, Y.; Colombel, F.; Himdi, M., "Meandered monopole coupled loop antenna," *Antennas and Propagation (EUCAP), 2012 6th European Conference on*, vol., no., pp. 3005,3008, 26-30 March 2012.

[29] Abdallah, M.; Le Coq, L.; Colombel, F.; Le Ray, G.; Himdi, M., "Frequency tunable monopole coupled loop antenna with broadside radiation pattern," *Electronics Letters*, vol.45, no.23, pp.1149,1151, November 2009.

Chapter 6

All-Dielectric Optical Nanoantennas

Alexandr E. Krasnok, Pavel A. Belov,
Andrey E. Miroshnichenko, Arseniy I. Kuznetsov,
Boris S. Luk'yanchuk and Yuri S. Kivshar

Additional information is available at the end of the chapter

1. Introduction

Antennas are important elements of wireless information communication technologies, along with sources of electromagnetic radiations and their detectors. One can say that antennas are at the heart of modern radio and microwave frequency communications technologies. They are at the front-ends of satellites, cell-phones, laptops and other communicating devices. In radio engineering, antennas refer to devices converting electric and magnetic currents into radio propagating waves and, vice versa, radio waves to currents. Recently, the concept of antennas have been extended to the optical frequency domain [1–9]

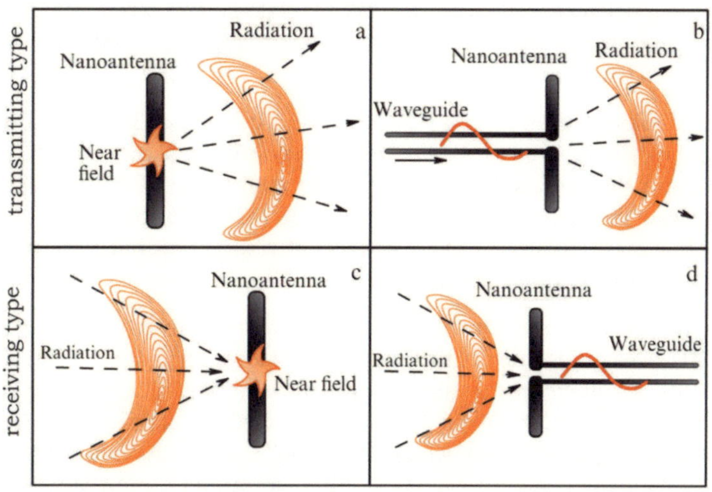

Figure 1. The basic principles of nanoantenna operation (exemplified by a nanodipole). Near field (a) or waveguide mode (b) transformation into freely propagating optical radiation; Panels (c, d) illustrate a reception regime. The configuration of feeding via a plasmonic waveguide is of great importance for practical applications of nanoantennas, especially for the development of wireless communication systems at the nanometer level, i.e., for future photonic chips. [8]

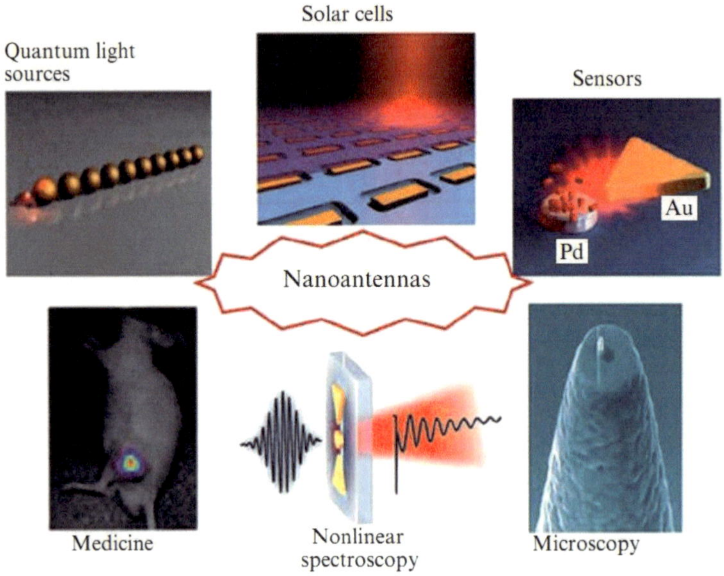

Figure 2. Plethora of nanoantennas application in modern science. [8]

as a result of the development of a new branch of physics emerged known as nanooptics, which studies the transmission and reception of optical signals by submicron and even nanometer-sized objects.

For nanooptics it is important to efficiently detect and direct the transmitting signals for optical information between nanoelements. The sources and detectors of radiation in nanooptics are nanoelements themselves, their clusters, and even individual molecules (atoms, ions). Nanoobjects functioning as antennas must exhibit high radiation efficiency and directivity.

Nanoantennas, similar to the radiofrequency antennas, are usually divided into two types, transmitting and receiving (see Fig. 1). Figure 1a schematically shows the interaction between a nanoantenna and the near field of an quantum emitter. In this case, the nanoantenna transforms the near field into freely propagating optical radiation, i.e. it is a transmitting nanoantenna. Figure 1c illustrates the operation of a receiving nanoantenna that converts incident radiation into a strongly confined near field.

The energy is usually delivered to a microwave antenna through a waveguide. Such an antenna converts waveguide modes to freely propagating radiation. In the case of optical antennas with their sufficiently small optical size, the waveguide mode must have the subwavelength cross section attainable by using so-called plasmonic waveguides. This type of nanoantenna feeding is depicted schematically in Fig. 1b. According to the reciprocity principle, such a nanoantenna is also capable of transforming incident radiation to plasmonic waveguide modes (see Fig. 1d).

Thus, the transmitting antenna converts a strongly confined field in the optical frequency range created by a certain (weakly emitting or almost non emitting) source into optical radiation (see Fig. 1a,b). Conversely, the receiving nanoantenna is a device efficiently converting incident light (optical frequency radiation) into a strongly confined field (see Fig. 1c,d), where an electromagnetic field is concentrated in a small region compared to the wavelength of light. Such fields are characterized by a spatial spectrum consisting mostly of evanescent waves. The confinement region may be of subwavelength dimension, leading to a strongly confined near field. The energy of this field contains contributions from stored and non radiated energy. However, an important particular case of nanoantennas is a device converting optical radiation into waveguide modes, and vice versa, as shown in Fig. 1c,d. In this case, the subwavelength dimension is characterized by the transverse cross section of the strongly confined field region. The longitudinal size of this region (along the waveguide axis) may be optically large, and the electromagnetic energy of the strongly confined field is referred to as expanding. The feeding configuration with a plasmonic waveguide is of great importance for practical applications of nanoantennas, especially for the development of wireless communication systems at the nanometer level, i.e., for future fully optical integrated circuits.

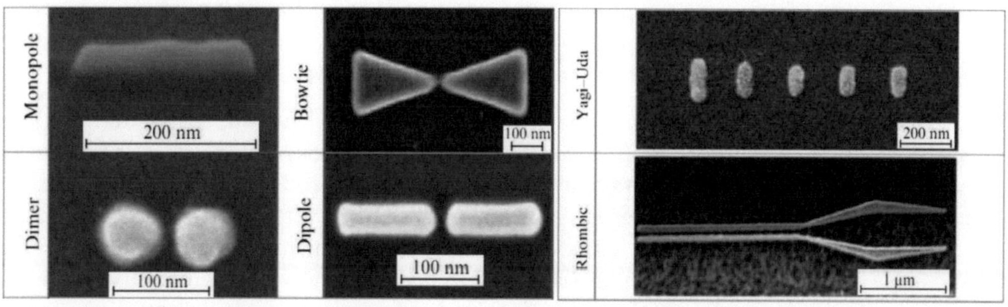

Figure 3. Main types of plasmonic nanoantennas. [8]

Nanoantennas are the most promising area of research in the modern nanooptics due to their ability to bridge the size and impedance mismatch between nanoemitters and free space radiation, as well as manipulate light on the scale smaller than the wavelength of light. Bearing in mind the great variety of sources and detectors of strongly confined optical fields (groups of atoms and molecules, luminescent and fluorescent cells, e.g., viruses and bacteria, sometimes individual molecules, quantum dots, and quantum wires), it is safe to say that the areas of practical applications of nanoantennas in the near future will be commensurate with that of their classical analogs.

At present, nanoantennas are used in near-field microscopy and high-resolution biomedical sensors; their application for hyperthermal therapy of skin neoplasms is a matter of the foreseeable future. There are some other potential applications of nanoantennas (see Fig. 2) that we believe to be equally promising, including solar cells [10], molecular and biomedical sensors [11], optical communication [12], and optical tweezers [13]. The variety of applications allows us to argue that the concept of nanoantennas presents a unique example of the penetration of new physics into various spheres of human activity.

Thus far, optical antennas have primarily been constructed from metallic materials, which support plasmonic resonances. The main types of plasmonic nanoantennas which have been realized experimentally are presented in Fig. 3. Different types of plasmonic nanoantennas are designed to perform various tasks. For example dipole nanoantennas [8, 9, 14–16] demonstrate high coefficient of electric field localization, while bowtie nanoantennas [8, 9, 13, 17–24] are broadband; Yagi-Uda type nanoantennas exhibit high directivity which is very useful for optical wireless communications on an optical chip [3, 8, 9, 12, 25–35]. However, despite of a number of advantages of plasmonic nanoantennas associated with their small size and strong localization of the electric field, such nanoantennas have large dissipative losses resulting in low radiation efficiency.

To overcome such limitations, we propose a new type of nanoantennas based on dielectric nanoparticles with a high index dielectric constant [8, 9, 36–44], for example Huygens optical elements and Yagi-Uda nanoantennas [see par.(3)]. Such all-dielectric optical nanoantennas will have low dissipative losses with enhanced magnetic response in the visible. The concept of optical magnetism based on dielectric nanoparticles is presented in the next section. The key for such novel functionalities of high index dielectric nanophotonic elements is the ability of subwavelength dielectric nanoparticles to support simultaneously both electric and magnetic resonances, which can be controlled independently. This type of nanoantennas has several unique features such as low optical losses at the nanoscale and superdirectivity. The concept of all-dielectric nanoantennas has been developed in our original papers [8, 9, 40–42, 45, 46] and also summarized below.

Furthermore all-dielectric nanoantennas allow us achieve the superdirectivity effect. Superdirectivity as a physical concept can be found in textbooks on antennas, however all so far proposed superdirective antennas are not reliably reproducible. More specifically, all previous attempts to achieve superdirectivity of antennas were based on discrete arrays of radiating dipoles with a rather cumbersome distribution of radiating currents over the array. This approach resulted in intrinsic drawbacks of known superdirective arrays - ultra-narrowfrequency range, high dissipation, and extreme sensitivity to any disturbance, etc. As a result, no single superdirective antenna was demonstrated up to now. In the context of nanoantennas, which originated from radio frequency antennas a few years ago, superdirectivity has never been discussed. However, superdirectivity would be a very desirable feature in nanophotonics with numerous useful applications. Here we describe [see par.(4)] the superdirectivity effect in a very simple, elegant, and practical way for a nanoparticle with a notch. This approach is able to shape higher-harmonics of the radiation field in such a way that not only superdirectivity of this nanoantennas becomes possible but also a strong subwavelength sensitivity of the radiation pattern to the location of the emitter can be easily realized.

2. Optical magnetism based on dielectric nanoparticles

It is well known that a pair of oscillating electric charges of opposite signs, know as an oscillating electric dipole, produces electromagnetic radiation at the oscillations frequency [48]. Although, distinct Şmagnetic chargesŤ, or monopoles, have not been observed so far, magnetic dipoles are very common sources of magnetic field in nature. The field of the magnetic dipole is usually calculated as the limit of a current loop shrinking to a point. Its profile is equivalent to the one of an electric dipole considering that the electric and magnetic fields are exchanged. The most common example of a magnetic dipole radiation is an electromagnetic wave produced by an excited metal split-ring resonator (SRR), which is a basic constituting element of metamaterials (see Fig. 4a) [49–57]. The real currents excited

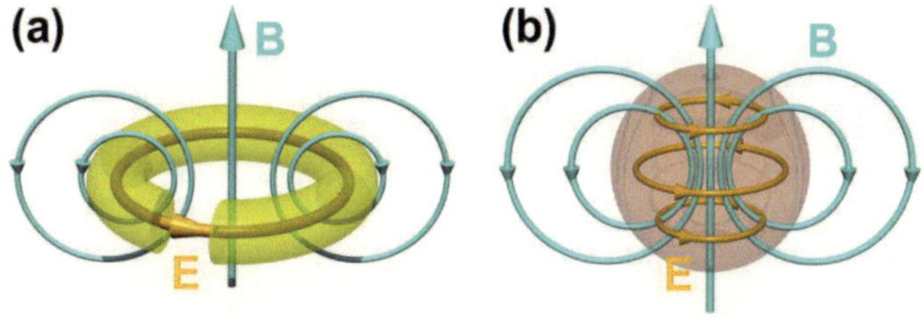

Figure 4. Schematic representation of electric and magnetic field distribution inside a metallic split-ring resonator (a) and a high-refractive index dielectric nanoparticle (b) at magnetic resonance wavelength. [47]

by external electromagnetic radiation and running inside the SRR produce a transverse oscillating up and down magnetic field in the center of the ring, which simulates an oscillating magnetic dipole. The major interest of these artificial systems is due to their ability to response to a magnetic component of incoming radiation and thus to have a non-unity or even negative magnetic permeability (μ) at optical frequencies, which does not exist in nature. This provides possibilities to design unusual material properties such as negative refraction [49–57], cloaking [58, 59], or superlensing [60]. The SRR concept works very well for gigahertz [55–57], terahertz [61, 62] and even near-infrared (few hundreds THz) [63–65] frequencies. However, for shorter wavelengths and in particular for visible spectral range this concept fails due to increasing losses and technological difficulties to fabricate smaller and smaller constituting split-ring elements [64, 66]. Several other designs based on metal nanostructures have been proposed to shift the magnetic resonance wavelength to the visible spectral range [49, 50]. However, all of them are suffering from losses inherent to metals at visible frequencies.

An alternative approach to achieve strong magnetic response with low losses is to use nanoparticles made of high-refractive index dielectric materials [53, 67]. As it follows from the exact Mie solution of light scattering by a spherical particle, there is a particular parameter range where strong magnetic dipole resonance can be achieved. Remarkably, for the refractive indices above a certain value there is a well-established hierarchy of magnetic and electric resonances. In contrast to plasmonic particles the first resonance of dielectric nanoparticles is a magnetic dipole resonance, and takes place when the wavelength of light inside the particle equals to the diameter $\lambda / n_S \simeq 2R_S$, where λ is a wavelength in a free space, R_S and n_S are the radius and refractive index of spherical particle. Under this condition the polarization of the electric field is anti-parallel at opposite boundaries of the sphere, which gives rise to strong coupling to circulation displacement currents while magnetic field oscillates up and down in the middle (see Fig. 4b).

Below in this section we present the experimental results demonstrating [47] that spherical silicon nanoparticles with sizes in the range from 100 nm to 200 nm have strong magnetic dipole response in the visible spectral range. The scattered ŞmagneticŤ light by these nanoparticles is so strong that it can be easily seen under a dark-field optical microscope. The wavelength of this magnetic resonance can be tuned throughout the whole visible spectral range from violet to red by just changing the nanoparticle size.

In article [47] we have chosen silicon (Si) as a material which has high refractive index in the visible spectral range (above 3.8 at 633 nm) on one side and still almost no dissipation losses on the other. Silicon nanorods have attracted considerable attention during the last few years

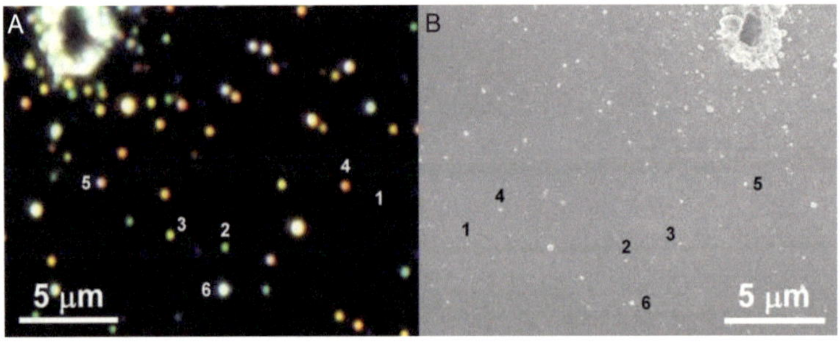

Figure 5. Dark-field microscope (a) and top-view scanning electron microscope (SEM) (b) images of the same area on a silicon wafer ablated by a femtosecond laser. Microscope image is inverted in horizontal direction relative to that of the SEM. Selected nanoparticles are marked by corresponding numbers 1 to 6 in both figures. [47]

due to their ability to change their visible color with the size [68]. This effect appears due to excitation of particular modes inside the cylindrical silicon nanoresonators. Moreover, recent theoretical work predicted that spherical silicon nanoparticles with sizes of a few/several hundred nanometers should have both strong magnetic and electric dipole resonances in the visible and near-IR spectral range [69, 70]. To fabricate the silicon nanoparticles we have used the laser ablation technique, which is an efficient method to produce nanoparticles of various materials and sizes [71]. Nanoparticles produced by the ablation method can be localized on a substrate and measured separately from each other using single nanoparticle spectroscopy.

Dark-field microscopic image of a silicon sample ablated by a focused femtosecond laser beam is shown in Fig. 5a. It shines by all the colours of the rainbow from violet to red. To clarify the origin of this strong scattering we selected some nanoobjects shining with different colours on the sample (see Fig. 5a) and measured their scattering spectra by single nanoparticle dark-field spectroscopy. Then, the same sample area was characterized by scanning electron microscopy and the selected nanoobjects providing different colours have been identified (see Fig. 5b, the dark-field microscope image is inverted in horizontal direction relative to that of the SEM). The results of this comparative analysis of the same nanoobjects by dark-field optical microscopy, dark-field scattering spectroscopy, and scanning electron microscopy are presented in Fig. 6. As it can be seen from the SEM images the observed colours are provided by silicon nanoparticles of almost perfect spherical shape and varied sizes. This makes it possible to analyze scattering properties of these nanoparticles in the frames of Mie theory [72] and identify the nature of optical resonances observed in our spectral measurements. The bottom panels (iv) in Fig. 6 represent a total extinction cross-section calculated using Mie theory [72] for silicon nanoparticles of different sizes (the calculations were done in free space). In these calculations, the size of the nanoparticles in each figure was chosen to be similar to the size defined from each corresponding SEM image (ii). It can be seen that there is a clear correlation between the experimental (iii) and theoretical spectra (iv) both in the number and position of the observed resonances. This makes it obvious that Mie theory describes more or less accurately our experimental results.

One of the main advantages of the analytical Mie solution compared to other computational methods is its ability to split the observed spectra into separate contributions of different multipole modes and have a clear picture of the field distribution inside the particle at each resonance maximum. This analysis was done for each particle size in Fig. 6 and corresponding multipole contributions were identified (see notations in the experimental and theoretical spectra). According to this analysis the first strongest resonance of these nanoparticles appearing in the longer wavelength part of the spectrum corresponds to magnetic dipole response (md). Electric field inside the particle at this resonance wavelength

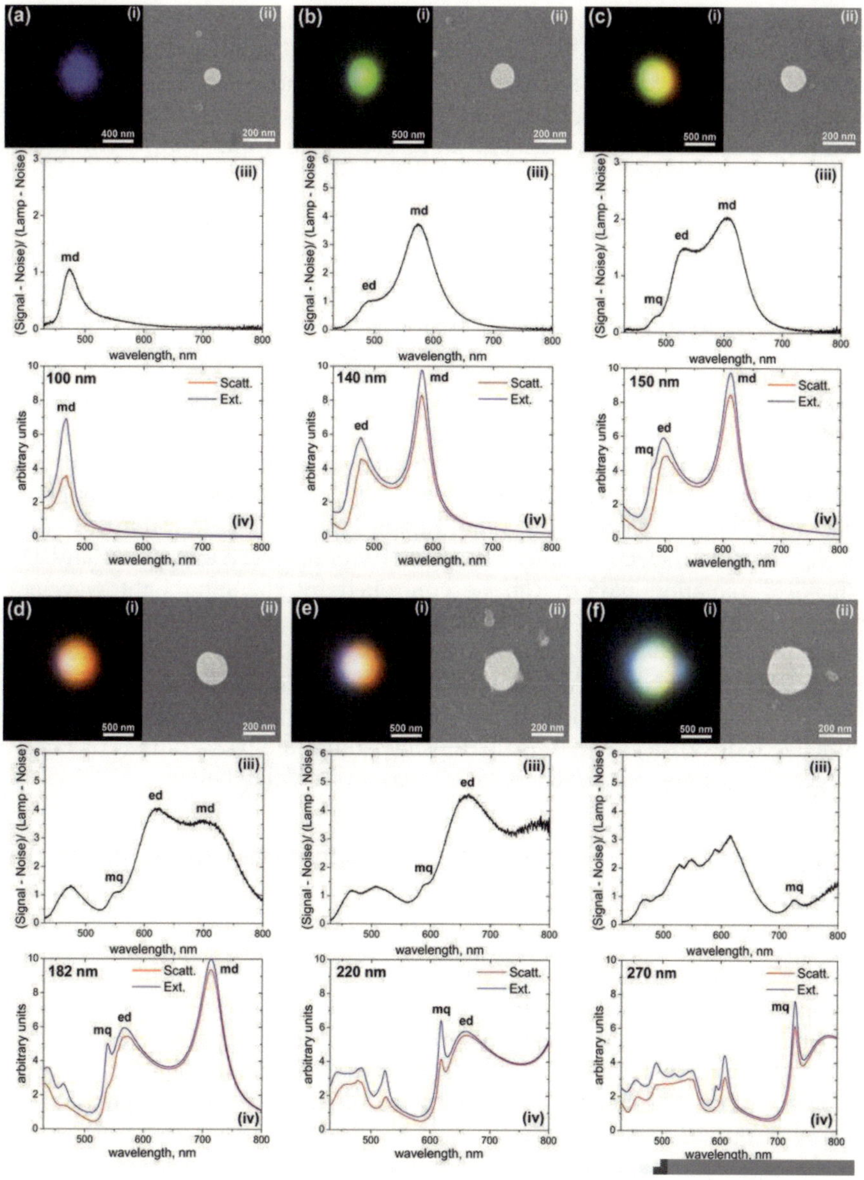

Figure 6. Close-view dark-field microscope (i) and SEM (ii) images of the single nanoparticles selected in Fig. 5. Figures (a) to (f) correspond to nanoparticles 1 to 6 from Fig. 5 respectively. (iii) Experimental dark-field scattering spectra of the nanoparticles. (iv) Theoretical scattering and extinction spectra calculated by Mie theory for spherical silicon nanoparticles of different sizes in free space. Corresponding nanoparticle sizes are defined from the SEM images (ii) and noted in each figure. [47]

has a ring shape while magnetic field oscillates in the particle center. Magnetic dipole resonance is the only peak observed for the smallest nanoparticles (see Fig. 6a). At increased nanoparticle size (see Fig. 6b,c) electric dipole (ed) resonance also appears at the blue part of the spectra, while magnetic dipole shifts to the red. For relatively small nanoparticles, the observed colour is mostly defined by the strongest resonance peak and changes from blue to green, yellow, and red when magnetic resonance wavelength shifts from 480 nm to 700 nm (see Fig. 6aŰd). So, we can conclude that the beautiful colours observed in the dark field microscope (see Fig. 5a) correspond to magnetic dipole scattering of the silicon nanoparticles, Şmagnetic lightŤ. Further increase of the nanoparticle size leads to the shift of magnetic and electric dipole resonances further to the red and infra-red frequencies, while higher multipole modes such as magnetic and electric quadrupoles appear in the blue part of the spectra (see Fig. 6dŰf).

Some differences between experimental and theoretical spectra observed in Fig. 6 can be attributed to the presence of silicon substrate, which is not taken into account in our simple Mie theory solution. We should also mention that very similar results have been published almost simultaneously by a different group of authors [73] who demonstrated magnetic and electric dipole resonances of silicon particles in red and near-IR spectral range.

Recently we have also experimentally demonstrated for the first time directional light scattering by spherical silicon nanoparticles in the visible spectral range [74]. These unique scattering properties arise due to simultaneous excitation and mutual interference of magnetic and electric dipole resonances inside a single nanosphere. This phenomenon is similar to a known since long time Kerker-type scattering predicted in [75] for hypothetical magneto-dielectric nanoparticles but never observed experimentally. Directivity of the far-field radiation pattern can be controlled by changing light wavelength and the nanoparticle size. Forward-to-backward scattering ratio above 6 was experimentally obtained at visible wavelengths. Similar directional light scattering by spherical ceramic particles in GHz [76] and GaAs nanodisks in the visible [77] has also been published almost simultaneously by different groups of authors. These unique optical properties of high-refractive index dielectric nanostructures constitute the background for our approach to all-dielectric nanoantennas, which will be discussed in detail below.

3. Huygens optical elements and YagiŰUda nanoantennas based on dielectric nanoparticles

Recently, it was suggested [8, 9, 40–42, 45, 46] a novel type of optical nanoantennas made of all-dielectric elements. Moreover, we argue that, since the source of electromagnetic radiation is applied externally, dielectric nanoantennas can be considered as the best alternative to their metallic counterparts. First, dielectric materials exhibit low loss at the optical frequencies. Second, as was suggested earlier, nanoparticles made of high-permittivity dielectrics may support both electric and magnetic resonant modes. This feature may greatly expand the applicability of optical nanoantennas for, e.g. for detection of magnetic dipole transitions of molecules [78]. In our study we concentrate on nanoparticles made of silicon. The real part of the permittivity of the silicon in the visible spectral range is about 16 [79], while the imaginary part is up to two orders of magnitude smaller than that of nobel metals (silver and gold).

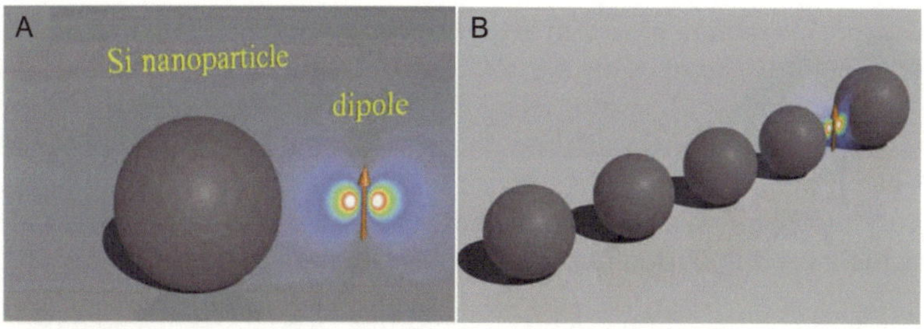

Figure 7. (**A**) Huygens element consisting of a single silicon nanoparticle and point-like dipole source separated by a distance $G_{ds} = 90$ nm (between dipole and sphere surface). The radius of the silicon nanoparticle is $R_s = 70$ nm. (**B**) Dielectric optical Yagi-Uda nanoantenna, consisting of the reflector of the radius $R_r = 75$ nm, and smaller director of the radii $R_d = 70$ nm. The dipole source is placed equally from the reflector and the first director surfaces at the distance G. The separation between surfaces of the neighbouring directors is also equal to G. [45]

3.1. General concept

The mentioned above properties of dielectric nanoparticles allow us to realize optical Huygens source [80] consisting of a point-like electric dipole operating at the magnetic resonance of a dielectric nanosphere (see Fig.7A). Such a structure exhibits high directivity with vanishing backward scattering and polarization independence, being attractive for efficient and compact designs of optical nanoantennas.

We start our analysis by considering a radiation pattern of two ideal coupled electric and magnetic dipoles. A single point-like dipole source generates the electric far-field of the following form

$$\mathbf{E}_p = \frac{k^2}{4\pi\epsilon_0 r} \exp(ikr) \left[\mathbf{p} - \mathbf{n}(\mathbf{n} \cdot \mathbf{p})\right], \tag{1}$$

where $\mathbf{p}$ is the electric dipole, $k = \omega/c$ is the wavenumber, $\mathbf{n}$ is the scattered direction, and r is the distance from the dipole source. The radiation pattern $\sigma = \lim_{r\to\infty} 4\pi r^2 |E_p|^2$ in the plane of the dipole $\mathbf{n} \times \mathbf{p} = 0$ is proportional to the standard figure-eight profile, $\sigma_{||} \propto |\cos\alpha|^2$, where α is the scattered angle. In the plane orthogonal to the dipole ($\mathbf{n} \cdot \mathbf{p} = 0$) the radiation pattern remains constant and angle independent, $\sigma_{\perp} \propto const$. Thus, the total radiation pattern of a single dipole emitter is a torus which radiates equally in the opposite directions. If we now place, in addition to the electric dipole, an orthogonal magnetic dipole located at the same point, the situation changes dramatically. The magnetic dipole $\mathbf{m}$ generates the electric far-field of the form

$$\mathbf{E}_m = -\sqrt{\frac{\mu_0}{\epsilon_0}} \frac{k^2}{4r\pi} \exp(ikr) \left(\mathbf{n} \times \mathbf{m}\right). \tag{2}$$

Thus, the total electric field is a sum of *two contributions* from both electric and magnetic dipoles $\mathbf{E}_{\text{total}} = \mathbf{E}_p + \mathbf{E}_m$. By assuming that the magnetic dipole is related to the electric dipole via the relation $|\mathbf{m}| = |\mathbf{p}|/(\mu_0\epsilon_0)^{1/2}$, which corresponds to an infinitesimally small wavefront of a plane wave often called a Huygens source [80], the radiation pattern becomes

$\sigma^H \propto |1 + \cos\alpha|^2$. This radiation pattern is quite different compared to that of a single electric dipole. It is *highly asymmetric* with the total suppression of the radiation in a particular direction, $\alpha = \pi$ [$\sigma^H(\pi) = 0$], and a strong enhancement in the opposite direction, $\alpha = 0$. The complete three-dimensional radiation pattern resembles a cardioid or apple-like shape, which is also azimuthally independent. Such a radiation pattern of the Huygens source is potentially very useful for various nanoantenna applications. However, while electric dipole sources are widely used in optics, magnetic dipoles are less common.

First, we consider an electric dipole source placed in a close proximity to a dielectric sphere [see Fig. 7(a)]. As was mentioned above, it can be analytically shown that high permittivity dielectric nanoparticles exhibit strong magnetic resonance in the visible range when the wavelength inside the nanoparticle equals its diameter $\lambda/n_s \approx 2R_s$ [81], where n_S and R_s are refractive index and radius of the nanoparticle, respectively. There are many dielectric materials with high enough real part of the permittivity and very low imaginary part, indicating low dissipative losses. To name just a few, silicon (Si, $\epsilon_1 = 16$), germanium (Ge, $\epsilon_1 = 20$), aluminum antimonide (AlSb, $\epsilon_1 = 12$), aluminum arsenide (AlAs, $\epsilon_1 = 10$), and other. In our study we concentrate on the nanoparticles made of silicon, which support strong magnetic resonance in the visible range for the radius varying from 40 nm to 80 nm [69].

For such a small radius compared to the wavelength $R_s < \lambda$, the radiation pattern of the silicon nanoparticle in the far field at the magnetic or electric resonances will resemble that of magnetic or electric point-like dipole, respectively. Moreover, it is even possible to introduce magnetic α^m and electric α^e polarisabilities [69, 72, 82] based on the Mie dipole scattering coefficients b_1 and a_1:

$$\alpha^e = \frac{6\pi a_1 i}{k^3}, \quad \alpha^m = \frac{6\pi b_1 i}{k^3}. \tag{3}$$

Thus, the dielectric nanoparticle excited by the electric dipole source at the magnetic resonance may result in the total far field radiation pattern which is similar to that of the Huygens source. Similar radiation patterns can be achieve in light scattering by a magnetic particle when permeability equals permittivity $\mu = \epsilon$, also known as Kerker's condition [75]. Our result suggests that even a dielectric nonmagnetic nanoparticle can support two induced dipoles of equal strength resulting in suppression of the radiation in the backward direction. Thus, it can be considered as the simplest and efficient optical nanoantenna with very good directivity.

In general, both polarisabilities α^m and α^e are nonzero in the optical region [69]. It is known that for a dipole radiation in the far field the electric and magnetic components should oscillate in phase to have nonzero energy flow. In the near field the electric and magnetic components oscillate with $\pi/2$ phase difference, thus, the averaged Poynting vector vanishes, and a part of energy is stored in the vicinity of the source. In the intermediate region, the phase between two components varies form $\pi/2$ to 0. Placing a nanoparticle close to the dipole source will change the phase difference between two components, and, thus, will affect the amount of radiation form the near field. In the case of plasmonic nanoparticles

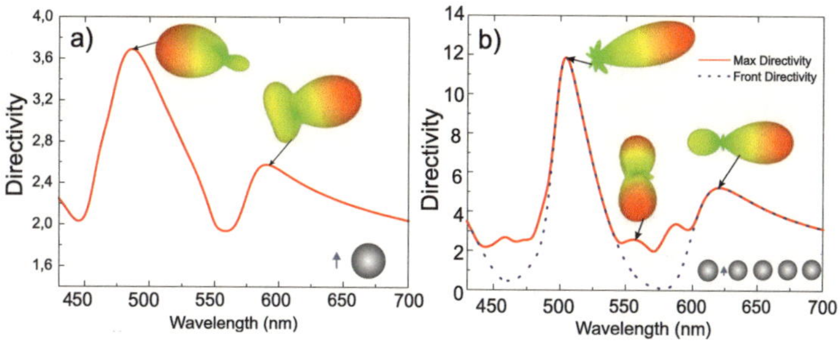

Figure 8. Wavelength dependence of the directivity of two types of all-dielectric nanoantennas consisting of (a) single dielectric nanoparticle of radius $R_d = 70$ nm, and (b) Yagi-Uda like design for the separation distance $D = 70$ nm. Insert shows 3D radiation pattern diagrams at particular wavelengths. [45]

which exhibit electric polarizability only, there is an abrupt phase change from 0 to π in the vicinity of the localized surface plasmon resonance, which makes it difficult to tune plasmonic nanoantennas for optimal performance. The dependence of the scattering diagram on the distance between the electric dipole source and metallic nanoparticle was studied in Ref. [83]. On contrary, in the case of nanoparticles with both electric and magnetic polarisabilities, it is possible to achieve more efficient radiation from the near to far field zone, due to subtle phase manipulation. *This is exactly the case of the dielectric nanoparticles.*

Any antenna is characterized by two specific properties, directivity (D) and radiation efficiency (η_{rad}), defined as [80, 84]

$$D = \frac{4\pi}{P_{\text{rad}}} \text{Max}[p(\theta, \varphi)], \quad \eta_{rad} = \frac{P_{\text{rad}}}{P_{\text{rad}} + P_{\text{loss}}}, \tag{4}$$

where P_{rad} and P_{loss} are integrated radiated and absorbed powers, respectively, θ and φ are spherical angles of standard spherical coordinate system, and $p(\theta, \varphi)$ is the radiated power in the given direction θ and/or φ. The directivity measures the power density of the antenna radiated in the direction of its strongest emission, while Radiation Efficiency measures the electrical losses that occur throughout the antenna at a given wavelength. To calculate these quantities numerically for the structures shown in Fig. 7a, we employ CST Microwave Studio. To get reliable results, we model the electric dipole source by a Discrete Port coupled to two PEC nanoparticles.

In Fig. 8(a) we show the dependence of the directivity on wavelength for a single dielectric nanoparticle excited by a electric dipole source. Two inserts demonstrate 3D angular distribution of the radiated pattern $p(\theta, \varphi)$ corresponding to the local maxima. In this case, the system radiates predominantly to the forward direction at $\lambda = 590$ nm, while in another case, the radiation is predominantly in the backward direction at $\lambda = 480$ nm. In this case, the total electric dipole moment of the sphere and point-like source and the magnetic dipole moment of the sphere oscillate with the phase difference $\arg(\alpha^m) - \arg(\alpha^e) = 1.3$rad, resulting in the destructive interference in the forward direction. At the wavelength $\lambda = 590$ nm the total electric and magnetic dipole moments oscillate in phase and produce Huygens-source-like radiation pattern with the main lobe directed in the forward direction. By adding more elements to the silicon nanoparticle, we can enhance the performance of all-dielectric nanoantennas. In particular, we consider a dielectric analogue of the Yagi-Uda design (see Fig.7) consisting of four directors and one reflector. The radii of the directors and the reflector are chosen to achieve the maximal constructive interference in the forward direction along the array. The optimal performance of the Yagi-Uda nanoantenna should be expected when the radii of the directors correspond to the magnetic resonance, and the radius of the reflector correspond to the electric resonance at a given frequency, with the coupling between the elements taken into account. Our particular design consists of the directors with radii $R_d = 70$ nm and the reflector with the radius $R_r = 75$ nm. In Fig. 8(b) we plot the directivity of all-dielectric Yagi-Uda nanoantenna vs. wavelength with the separation distance $D = 70$ nm. Inserts demonstrate the 3D radiation patterns at particular wavelengths. We achieve a strong maximum at $\lambda = 500$ nm. The main lobe is extremely narrow with the beam-width about 40° and negligible backscattering. The maximum does not correspond exactly to either magnetic or electric resonances of a single dielectric sphere, which implies the importance of the interaction between constitutive nanoparticles.

As the next step, we study the performance of the all-dielectric nanoantennas for different separation distances D, and compare it with *a plasmonic analogue* of the similar geometric design made of silver nanoparticles. According to the results summarized in Fig. 9, the radiation efficiencies of both types of nanoantennas *are nearly the same* for larger separation of directors $D = 70$ nm with the averaged value 70%. Although dissipation losses of silicon are much smaller than those of silver, the dielectric particle absorbs the EM energy by the whole spherical volume, while the metallic particles absorb mostly at the surface. As a result, there is no big difference in the overall performance of these two types of nanoantennas for relatively large distances between the elements. However, the difference becomes *very strong* for smaller separations. The radiation efficiency of the all-dielectric nanoantenna is insensitive to the separation distance [see Fig. 9 (a)]. On contrary, the radiation efficiency drops significantly for metallic nanoantennas [see Fig. 9 (b)].

Finally, we investigate the modification of the transition rate of a quantum point-like source placed in the vicinity of dielectric particles. For electric-dipole transitions and in the weak-coupling regime, the normalised spontaneous decay rate Γ/Γ_0, also known as Purcell factor, can be calculated classically as the ratio of energy dissipation rates of an electric dipole P/P_0 [7]. Here, Γ_0 and P_0 correspond to transition rate of the quantum emitter and energy dissipation rate of the electric dipole in free space [85]. In the limit of the intrinsic quantum yield of the emitter close to unity, both ratios become equal to each other $\Gamma/\Gamma_0 = P/P_0$, which allows us to calculate the Purcell factor in the classical regime [7]. We have calculated the Purcell factor by using both, numerical and analytical approaches. Numerically, by using

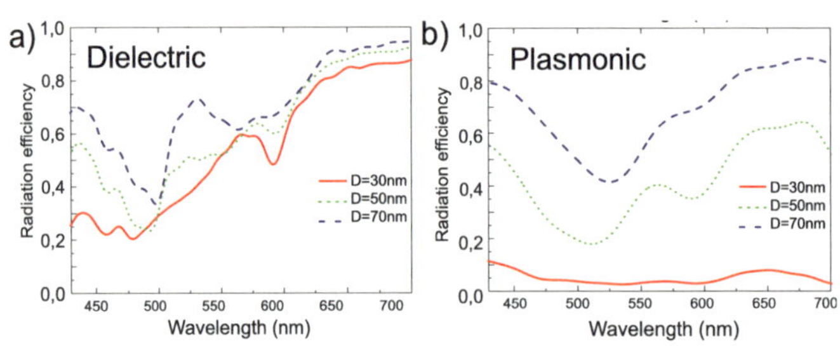

Figure 9. Radiation efficiencies of (a) dielectric (Si) and (b) plasmonic (Ag) Yagi-Uda optical nanoantennas of the same geometrical designs for various values of the separation distance D. [45]

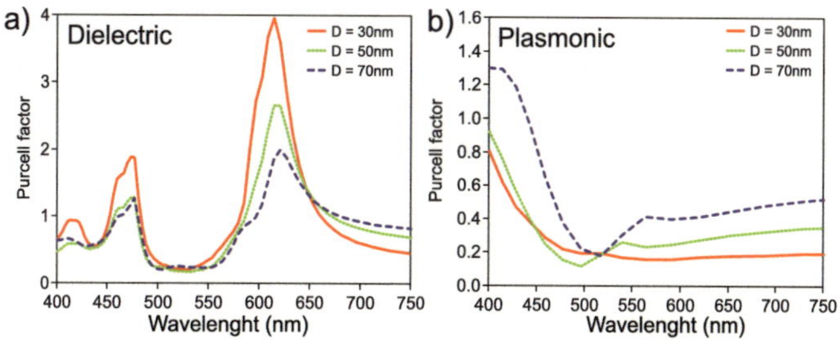

Figure 10. Purcell factor of all-dielectric Yagi-Uda nanoantenna vs wavelength for various values of the separation distance *D*. [45]

the CST Microwave Studio we calculate the total radiated in the far-field and dissipated into the particles powers and take the ratio of their sum to the total power radiated by the electric dipole in free space. Analytically, we employed the generalised multiparticle Mie solution [86] adapted for the electric dipole excitation [87]. We verified that both approaches produce similar results. In Fig. 10 we show calculated Purcell factor of the all-dielectric Yagi-Uda nanoantenna vs. wavelength for various separation distances. We observe that, by decreasing the separation between the directors, the Purcell factor becomes stronger near the magnetic dipole resonance. We can notice that a plasmonic analogue of the same nanoantenna made of Ag exhibits low Purcell factor less than one. Thus, such relatively high Purcell factor can be employed for efficient photon extraction from molecules placed near all-dielectric optical nanoantennas.

3.2. Experimental verification of dielectric Yagi-Uda nanoantenna

There are exist some technological issues to reproduce an object of the nanometer size with a high accuracy. For this reason we have scaled the dimensions of the proposed optical all-dielectric Yagi-Uda nanoantenna to the microwave frequency range while keeping all the material parameters in order to study the microwave analogue of the nanoantenna experimentally. We use the design of the Yagi-Uda antenna shown in Fig. 7b. To mimic the silicon spheres in microwave frequency range, we employ MgO-TiO_2 ceramic which is characterized by dielectric constant of 16 and dielectric loss factor of $(1.12-1.17)10^{-4}$ measured at frequency 9-12 GHz [88]. As a source, we use a half-wavelength vibrator. We study experimentally both the radiation pattern and directivity of the antenna.

We set the radius of the reflector equal to $R_r = 5$ mm. The frequencies of the electric and magnetic Mie resonances of the sphere calculated with the help of Eq. (3) are 10.2 GHz and 7 GHz, respectively. The radius of the directors is $R_d = 4$ mm. In this case, the frequencies of the electric and magnetic Mie resonances are 12.5 GHz and 9 GHz. As a source, we model a half-wavelength vibrator with the total length of $L_v = 19.8$ mm and diameter of $D_v = 2.2$ mm. The distances between the reflector, directors, and vibrator have been adjusted by numerical simulations. We achieve an effective suppression of the back and minor lobes, and the narrow major lobe (of about 40°) of the antenna when the distance between the director's surface as well as the distance between vibrator center and the first director surface are 1.5 mm; the distance between the surface of the reflector and vibrator centre is 1.1 mm.

Figures 11(a,b) show the photographs of the fabricated all-dielectric Yagi-Uda antenna. The reflector and directors are made of MgO-TiO_2 ceramic with accuracy of ± 0.05 mm. To

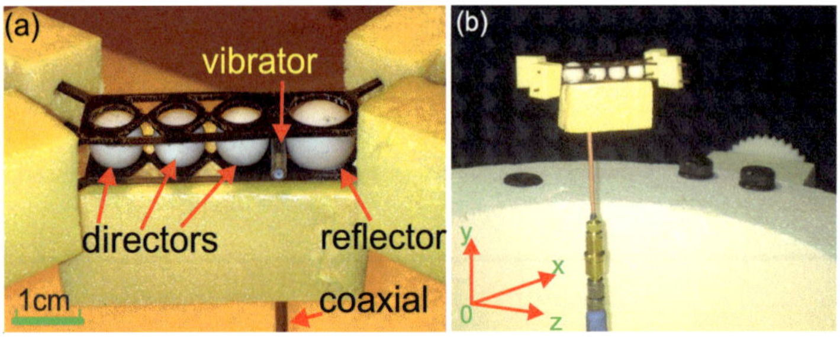

Figure 11. Photographs of the all-dielectric Yagi-Uda microwave antenna. (a) Detailed view of the antenna placed in a holder. (b) Antenna placed in an anechoic chamber; the coordinate z is directed along the vibrator axis; the coordinate y is directed along the antenna axis. [46]

fasten together the elements of the antenna and vibrator, we use a special holder made of a thin dielectric substrate with dielectric permittivity close to 1 [being shown in Fig. 11(a)]. Styrofoam material with the dielectric permittivity of 1 is used to fix the antenna in the azimuthal-rotation unit [see Fig. 11(b)]. To feed the vibrator, we employ a coaxial cable that is connected to an Agilent PNA E8362C vector network analyzer.

Any antenna is characterized by the total directivity (4). Sometimes it is not possible to determine the value of the total directivity experimentally due to difficulties to measure the total radiated power P_{rad}. In this case, it is convenient to use directivity in the planes where electric field **E** and magnetic field **H** oscillate in the far field. For our coordinates the directivity in the evaluation plane (E-plane) and the azimuthal plane (H-plane) can be expressed as:

$$D_E = \left.\frac{2\pi \text{Max}[p(\theta)]}{\int_0^{2\pi} p(\theta) d\theta}\right|_{\varphi=0}, \quad D_H = \left.\frac{2\pi \text{Max}[p(\varphi)]}{\int_0^{2\pi} p(\varphi) d\varphi}\right|_{\theta=\pi/2}. \tag{5}$$

Equations (5) are multiplied by 2π because of the integration in the denominator is performed only for one coordinate while the second coordinate is fixed.

To extract the antenna directivity in the E- and H-planes from the experimental data, we measure the radiated power by the antenna in the frequency range from 10 GHz to 12 GHz with a step of 50 MHz. Then, by employing Eq. (5) we calculate the directivity at each frequency. The results are presented in Fig. 12a. To estimate the performance of

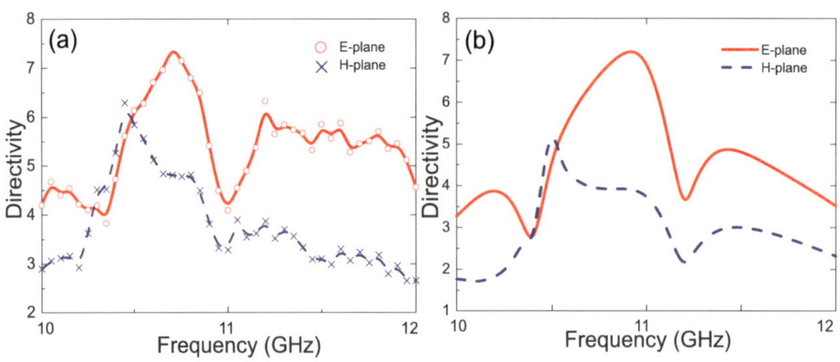

Figure 12. (a) Experimentally measured and (b) numerically calculated antenna's directivity in both E- and H-planes. [46]

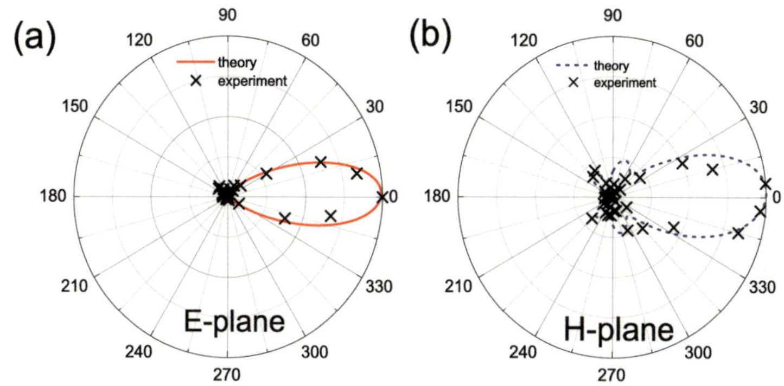

Figure 13. Radiation pattern of the antenna in (a) E-plane and (b) H-plane at the frequency 10.7 GHz. Solid lines show the result of numerical simulations in CST; the crosses correspond to the experimental data. [46]

the all-dielectric Yagi-Uda antenna at microwaves, we simulate numerically the antenna's response by employing the CST Microwave Studio. We observe excellent agreement between numerical results of Fig. 12b and measured experimental data. However, we notice a small frequency shift of the measured directivity (approx. 2%) in comparison with the numerical results. This discrepancies can be explained by the effect of the antenna holders in the experiment, not included into the numerical simulation.

The antenna radiation patterns in the far field (at the distance $\simeq$ 3 m) are measured in an anechoic chamber by a horn antenna and rotating table. The measured radiation patterns of the antenna in E- and H-planes at the frequency 10.7 GHz are shown in Fig. 13. The measured characteristics agree very well with the numerical results. A small disagreement can be explained by the presence of the antenna holder which influence was not taken into account in our numerical simulations.

4. All–dielectric superdirective optical nanoantenna

For optical wireless circuits on a chip, nanoantennas are required to be both highly directive and compact [12, 89–91]. In nanophotonics, directivity has been achieved for arrayed plasmonic antennas utilizing the Yagi-Uda design [37, 84, 90, 92, 93], large dielectric spheres [94], and metascreen antennas [95]. Though individual elements of these arrays are optically small, the overall size of the radiating systems is larger than the radiation wavelength λ. In addition, small plasmonic nanoantennas possess weak directivity close to the directivity of a point dipole [90, 96, 97].

As was discussed above, it was suggested theoretically and experimentally to employ magnetic resonances of high-index dielectric nanoparticles for enhancing the nanoantenna directivity [8, 9, 37, 40–42, 45, 46, 98]. High-permittivity nanoparticles can have nearly resonant balanced electric and magnetic dipole responses. This balance of the electric and magnetic dipoles oscillating with the same phase allows the practical realization of the Huygens source, an elementary emitting system with a cardioid pattern [37, 44, 46, 80] and with the directivity larger than 3.5. Importantly, a possibility to excite magnetic resonances leads to the improved nanoantenna directional properties without a significant increase of its size.

Superdirectivity has been already discussed for radio-frequency antennas, and it is defined as directivity of an electrically small radiating system that significantly exceeds (at least in 3 times) directivity of an electric dipole [80, 99, 100]. In that sense, the Huygens source is not superdirective. In the antenna literature, superdirectivity is claimed to be achievable only in antenna arrays by the price of ultimately narrow frequency range and by employing very precise phase shifters (see, e.g., Ref. [80, 99, 100]). Therefore, superdirective antennas, though very desirable for many applications such as space communications and radioastronomy, were never demonstrated and implemented for practical applications.

Superdirectivity was predicted theoretically for an antenna system [95] where some phase shifts were required between radiating elements to achieve complex shapes of the elements of a radiating system which operates as an antenna array. In this paper, we employ the properties of subwavelength particles excited by an inhomogeneous field with higher-order magnetic multipoles. We consider a subwavelength dielectric nanoantenna (with the size of 0.4 wavelength) with a notch resonator excited by a point-like emitter located in the notch. The notch transforms the energy of the generated magneto-dipole Mie resonance into high-order multipole moments, where the magnetic multipoles dominate. This system is resonantly scattering i.e. it is very different from dielectric lenses and usual dielectric cavities which are large compared to the wavelength. Another important feature of the notched resonator is its huge sensitivity of the radiation direction to a spatial position of the emitter. This property leads to a strong beam steering effect and subwavelength sensitivity of the radiation direction to the source location. The proposed design of superdirective nanoantennas may also be useful for collecting single-source radiation, monitoring quantum objects states, and nanoscale microscopy. In order to achieve superdirectivity, we should generate subwavelength spatial oscillations of the radiating currents [80, 99, 100]. Then, near fields of the antenna become strongly inhomogeneous, and the near-field zone expands farther than that of a point dipole. The effective antenna aperture can be defined as $S = D_{max}\lambda^2/(4\pi)$, where the maximum of directivity $D_{max} = 4\pi P_{max}/P_{tot}$, λ is the wavelength in free space in our case, P_{max} and P_{tot} are respectively the maximum power in the direction of the radiation pattern and the total radiation power. By normalizing the effective aperture S by the geometric aperture for a spherical antenna $S_0 = \pi R_S^2$, we obtain the definition of superdirectivity [80, 99]:

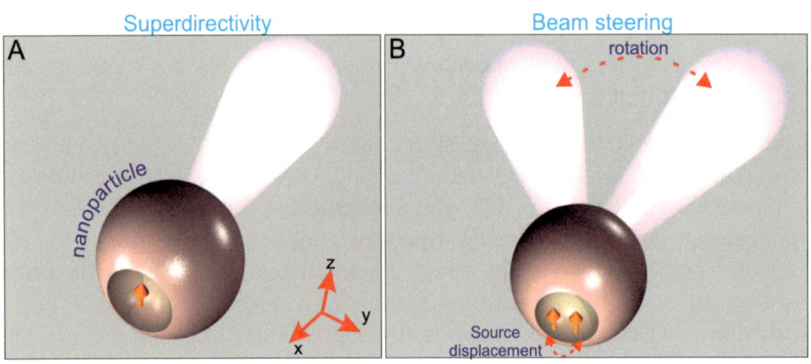

Figure 14. (**A**) Geometry of an all-dielectric superdirective nanoantenna excited by a point-like dipole. (**B**) Concept of the beam steering effect at the nanoscale.

$$S_n = \frac{D_{max}\lambda^2}{4\pi^2 R_S^2} \gg 1 \quad (6)$$

Practically, the value $S_n = 4 \ldots 5$ is sufficient for superdirectivity of a sphere. In this work, maximum of 6.5 for S_n is predicted theoretically for the optical frequency range, and the value of 5.9 is demonstrated experimentally for the microwave frequency range.

4.1. Concept of all–dielectric superdirective optical nanoantennas

Here we demonstrate a possibility to create a superdirective nanoantenna without hypothetic metamaterials and plasmonic arrays. We consider a silicon nanoparticle, taking into account the frequency dispersion of the dielectric permittivity [79]. The radius of the silicon sphere is equal in our example to $R_S = 90$ nm. For a simple sphere under rather homogeneous (e.g. plane-wave) excitation, only electric and magnetic dipoles can be resonantly excited while the contribution of higher-order multipoles is negligible in the visible [37]. Making a notch in the sphere breaks the symmetry and increases the contribution of higher-order multipoles into scattering even if the sphere is still excited homogeneously. Further, placing a nanoemitter (e.g. a quantum dot) inside the notch, as shown in Fig. 14 we create the conditions for the resonant excitation of multipoles: the field exciting the resonator is now spatially very non-uniform as well as the field of a set of multipoles. In principle, the notched particle operating as a nanoantenna can be performed by different semiconductor materials and have various shapes – spherical, ellipsoidal, cubic, conical, as well as the notch. However, in this work, the particle is a silicon sphere and the notch has the shape of a hemisphere with a radius $R_n < R_S$. The emitter is modeled as a point-like dipole and it is shown in Fig. 14 by a red arrow.

It is important to mention that our approach is seemingly close to the idea of references [101, 102] where a small notch on a surface of a semiconductor microlaser was used to achieve higher emission directivity by modifying the field distribution inside the resonator [103]. An important difference between those earlier studies and our work is that the design discussed earlier is not optically small and the directive emission is not related to superdirectivity. In our case, the nanoparticle is much smaller than the wavelength, and our design allows superdirectivity. For the same reason our nanoantenna is not dielectric [104, 105] or Luneburg [106, 107] lenses. For example, immersion lenses [108–111] are the smallest from known dielectric lenses, characterized by the large size 1-2 μm in optical frequency range. The working methodology of such lenses is to collect a radiation by large geometric aperture S, while $S_n \simeq 1$. *Our approach demonstrates that the subwavelength system, with small geometric aperture, can have high directing power because of an increase of the effective aperture.* Moreover, there are articles (see. references [85, 112]) where the transition rates of atoms inside and outside big dielectric spheres with low dielectric constant (approximately 2), were studied.

First, we consider a particle without a notch but excited inhomogeneously by an emitter point. To study the problem numerically, we employed the simulation software CST Microwave Studio. Image Fig. 15A shows the dependence of the maximum directivity D_{max} on the position of the source in the case of a sphere $R_S = 90$ nm without a notch, at the wavelength λ = 455 nm (blue curve with crosses). This dependence has the maximum (D_{max} = 7.1) when the emitter is placed inside the particle at the distance 20 nm from its surface. The analysis shows that in this case the electric field distribution inside a particle corresponds to the noticeable excitation of higher-order multipole modes not achievable with the homogeneous excitation.

Furthermore, the amplitudes of high-order multipoles are significantly enhanced with a small notch around the emitter, as it is shown in Fig. 14. This geometry transforms it into a resonator with high-order multipole moments. In this example the center of the notch is on the nanosphere's surface. The optimal radius of the notch (for maximal directivity) is R_n = 40 nm. In Fig. 15A the extrapolation red curve with circles, corresponding to simulation results, shows the maximal directivity versus the location of the emitter at the wavelength 455 nm. The Fig. 15B shows the directivity versus λ with and without a notch, it exhibits a maximum of 10 for the directivity at 455 nm. The inset shows the three-dimensional radiation pattern of the structure at λ =455 nm. This pattern has an angular width (at the level of 3 dB) of the main lobe equal to 40°. This value of directivity corresponds to the normalized effective aperture S_n = 6.5.

Figures Fig. 16A and B show the distribution of the absolute values and phases of the internal electric field in the vicinity of the nanoantenna. Electric and magnetic fields inside the particle are strongly inhomogeneous at λ =455 nm i.e. in the regime of the maximal directivity. In this regime, the internal area where the electric field oscillates with approximately the same phase turns out to be maximal. This area is located near the back side of the spherical particle, as can be seen in figure Fig. 16B,D. In other words, the effective near zone of the nanoantenna is maximal in the superdirective regime.

Usually, high directivity of plasmonic nanoantennas is achieved by the excitation of higher *electrical* multipole moments in plasmonic nanoparticles [83, 113, 114] or for core-shell resonators consisting of a plasmonic material and a hypothetic metamaterial which would demonstrate the extreme material properties in the nanoscale [115]. Although, the values of directivity achieved for such nanoantennas do not allow superdirectivity, these studies stress the importance of higher multipoles for the antenna directivity.

Next, we demonstrate how to find multipole modes excited in the all-dielectric superdirective nanoantenna which are responsible for its enhanced directivity. We expand the exactly

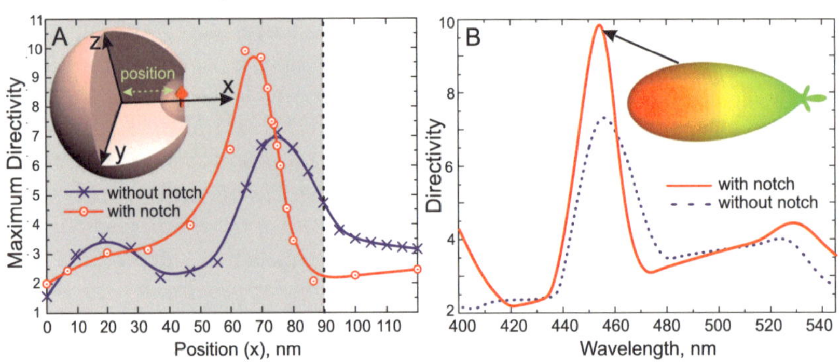

Figure 15. (**A**) Maximum of directivity depending on the position of the emitter (λ = 455 nm) in the case of a sphere with and without notch. Vertical dashed line marks the particle radius centered at the coordinate system. (**B**) Directivity dependence on the radiation wavelength. The inset shows three-dimensional radiation pattern of the structure (R_s = 90 nm and R_n = 40 nm).

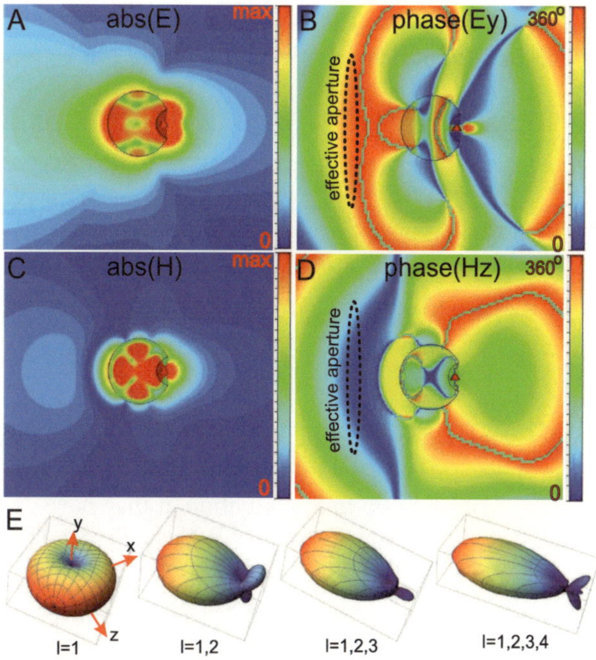

Figure 16. Distribution of (**A**) absolute values and (**B**) phases of the electric field (**C** and **D** for magnetic field, respectively) of the all-dielectric superdirective nanoantenna with source in the center of notch, at the wavelength $\lambda = 455$ nm. (**E**) Dependence of the radiation pattern of all-dielectric superdirective nanoantenna on the number of taken into account multipoles. Dipole like source located along the z axis.

simulated internal field, producing the polarization currents in the nanoparticle, into multipole moments following to [116]. The expansion is a series of vector spherical harmonics with the coefficients $a_E(l,m)$ and $a_M(l,m)$, which characterize the electrical and magnetic multipole moments [116]:

$$a_E(l,m) = \frac{4\pi k^2}{i\sqrt{l(l+1)}} \int Y^*_{lm} \left[\rho \frac{\partial}{\partial r}[r j_l(kr)] + \frac{ik}{c}(\mathbf{r}\cdot\mathbf{j}) j_l(kr)\right] d^3x,$$

$$a_M(l,m) = \frac{4\pi k^2}{i\sqrt{l(l+1)}} \int Y^*_{lm} \mathrm{div}\left(\frac{\mathbf{r}\times\mathbf{j}}{c}\right) j_l(kr) d^3x, \tag{7}$$

where $\rho = 1/(4\pi)\mathrm{div(E)}$ and $\mathbf{j} = c/(4\pi)(\mathrm{rot(H)} + ik\mathrm{E})$ are densities of the *total* electrical charges and currents that can be easily expressed through the internal electric E and magnetic H fields of the sphere, Y_{lm} are the spherical harmonics of the orders ($l > 0$ and $0 \geq |m| \leq l$), $k = 2\pi/\lambda$, $j_l(kr)$ are the l-order spherical Bessel function and c is the speed of light. Coefficients $a_E(l,m)$ and $a_M(l,m)$ determine the electric and magnetic mutipole moments, namely dipole at $l = 1$, quadrupole at $l = 2$, octupole at $l = 3$ etc.

The multipole coefficients determine not only the mode structure of the internal field but also the angular distribution of the radiation. In particular, in the far field zone electric and magnetic fields of l-order multipole depend on the distance r as [116] $\sim (-1)^{i+1}\frac{\exp(ikr)}{kr}$ and expression for the angular distribution of the radiation power can be written as follows:

$$\frac{dP(\theta,\varphi)}{d\Omega} = \frac{c}{8\pi k^2}\left|\sum_{l,m}(-i)^{l+1}[a_E(l,m)\mathbf{X}_{lm}\times\mathbf{n}+a_M(l,m)\mathbf{X}_{lm}]\right|^2,$$

$$\mathbf{X}_{lm}(\theta,\varphi) = \frac{1}{\sqrt{l(l+1)}}\begin{bmatrix} A^-_{l,m}Y_{l,m+1}+A^+_{l,m}Y_{l,m-1} \\ -iA^-_{l,m}Y_{l,m+1}+iA^+_{l,m}Y_{l,m-1} \\ mY_{l,m}\end{bmatrix}, \qquad (8)$$

where $A^{\pm}_{l,m} = (1/2)\sqrt{(l\pm m)(l\mp m+1)}$, $d\Omega = \sin(\theta)d\theta d\varphi$ is the solid angle element in spherical coordinates and $\mathbf{n}$ - unit vector of the observation point. All coefficients $a_E(l,m)$ and $a_M(l,m)$ give the same contribution to the radiation, if they have the same values. Since higher-order multipoles for optically small systems have usually negligibly small amplitudes compared to $a_E(1,m)$ and $a_M(1,m)$, they are, as a rule, not considered.

The amplitudes of multipole moments, are found by using the expressions (7) for electric and magnetic fields distribution Fig. 16A-D are shown in Fig. 17, where we observe strong excitation of $a_E(1,0)$, $a_M(1,1)$, $a_M(1,-1)$, $a_M(2,2)$, $a_M(2,-2)$, $a_M(3,3)$, $a_M(3,-3)$, $a_M(4,2)$, $a_M(4,-2)$, $a_M(4,4)$ and $a_M(4,-4)$. These multipole moments determine the angular pattern of the antenna. All other ones give a negligible contribution. Absolute values of all magnetic moments are larger than those of the electric moments in the corresponding multipole orders, and the effective spectrum of magnetic multipoles is also broader than the one of the electric moments. Thus, the operation of the antenna is mainly determined by the magnetic multipole response. Absolute values of multipole coefficients $a_M(l,\pm|m|)$ of the same order l are practically equivalent. However, the phase of some coefficients are different. Therefore, the modes with $+|m|$ and $-|m|$ form a strong anisotropy of the forward–backward directions that results in the unidirectional radiation.

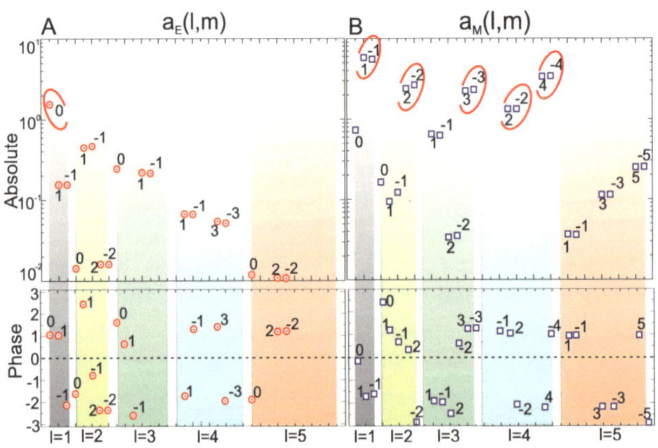

Figure 17. Absolute values and phases of (**A**) electric and (**B**) magnetic multipole moments that provide the main contribution of the radiation of all-dielectric superdirective optical nanoantenna at the wavelength 455 nm. Multipole coefficients providing the largest contribution to the antenna direction are highlighted by red circles.

We have performed the transformation of multipole coefficients into an angular distribution of radiation in accordance to (8) by using distribution of the electric and magnetic fields Fig. 16A-D and determined the relative contribution of each order l. Fig. 16E shows how the directivity grows versus the spectrum of multipoles with equivalent amplitudes. The right panel of Fig. 16E nearly corresponds to the inset in Fig. 15 that fits to the results shown in Fig. 16E.

Generally, the superdirectivity effect is accompanied by a significant increase of the effective near field zone of the antenna compared to the one of a point dipole for which the near zone radius is equal to $\lambda/2\pi$. In the optical frequency range this effect is especially important, considering the crucial role of the near fields at the nanoscale.

Usually, the superdirectivity regime corresponds to a strong increase of dissipative losses [80]. Radiation efficiency of the nanoantenna is determined by $\eta_{rad} = P_{rad}/P_{in}$, where P_{in} is the accepted input power of the nanoantenna. However, the multipole moments excited in our nanoantenna are mainly of magnetic type that leads to a strong increase of the near magnetic field that dominates over the electric one. Since the dielectric material does not dissipate the magnetic energy, the effect of superdirectivity does not lead to a so large increase of losses in our nanoantenna as it would be in the case of dominating electric multipoles. However, since the electric near field is nonzero the losses are not negligible. At wavelengths 440-460 nm (blue light) the directivity achieves 10 but the radiation efficiency is less than 0.1 (see [Fig. 18)]. This is because silicon has very high losses in this range [79]. Peak of directivity is shifted to longer wavelengths with the increase of the nanoantenna size. For the design parameters corresponding to the operation wavelength 630 nm (red light) the calculated value of radiation efficiency is as high as 0.5, with nearly same directivity close to 10. In the infrared range, there are high dielectric permittivity materials with even lower losses. In principle, the proposed superdirectivity effect is not achieved by price of increased losses, and this is an important advantage compared to known superdirective radio-frequency antenna arrays [80] and compared to their possible optical analogues – arrays of plasmonic nanoantennas.

4.2. Steering of light at the nanoscale

Here we examine the response of the nanoantenna to subwavelength displacements of the emitter. Displacement in the plane perpendicular to the axial symmetry of antenna (i.e. along

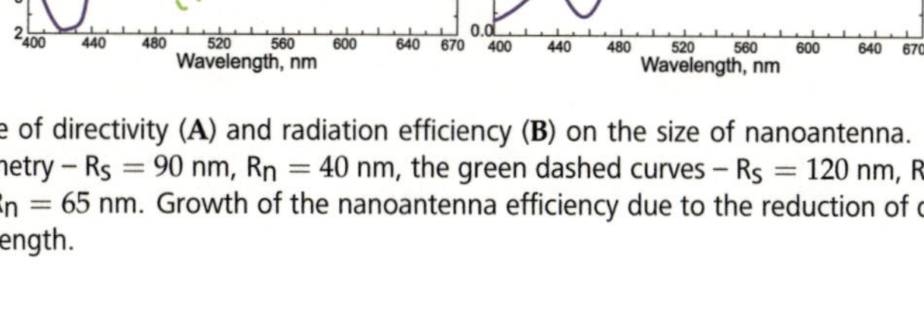

Figure 18. Dependence of directivity (**A**) and radiation efficiency (**B**) on the size of nanoantenna. Here, the blue solid lines corresponds to the geometry – $R_s = 90$ nm, $R_n = 40$ nm, the green dashed curves – $R_s = 120$ nm, $R_n = 55$ nm and red point curves – $R_s = 150$ nm, $R_n = 65$ nm. Growth of the nanoantenna efficiency due to the reduction of dissipative losses in silicon with increasing of wavelength.

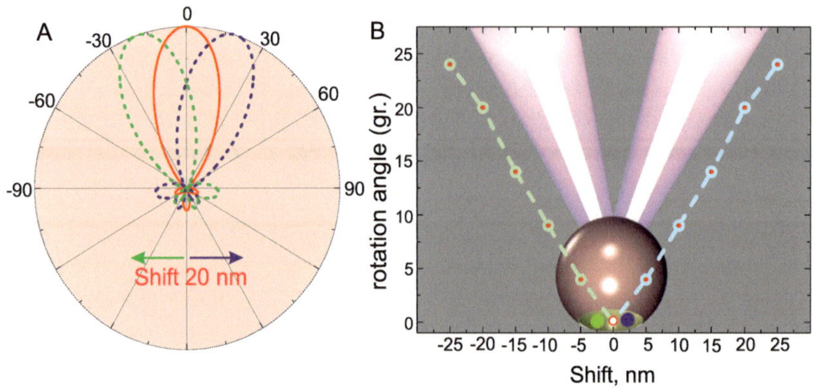

Figure 19. The rotation effect of the main beam radiation pattern, with subwavelength displacement of emitter inside the notch. (**A**) The radiation patterns of the antenna with the source in center (solid line) and the rotation of the beam radiation pattern for the 20 nm left/right offset (dashed lines). (**B**) Dependence of the rotation angle on the source offset.

the y axis) leads to the rotation of the beam *without damaging the superdirectivity*. Fig.19A shows the radiation patterns of the antenna with the source in center (solid line) and the rotation of the beam for the 20 nm left/right offset (dashed lines). Shifting of the source in the right side leads to the rotation of pattern to the left, and vice versa. The angle of the beam rotation is equal to 20 degrees, that is essential and available to experimental observations. The result depends on the geometry of the notch. For a hemispherical notch, the dependence of the rotation angle on the displacement is presented in Fig.19B.

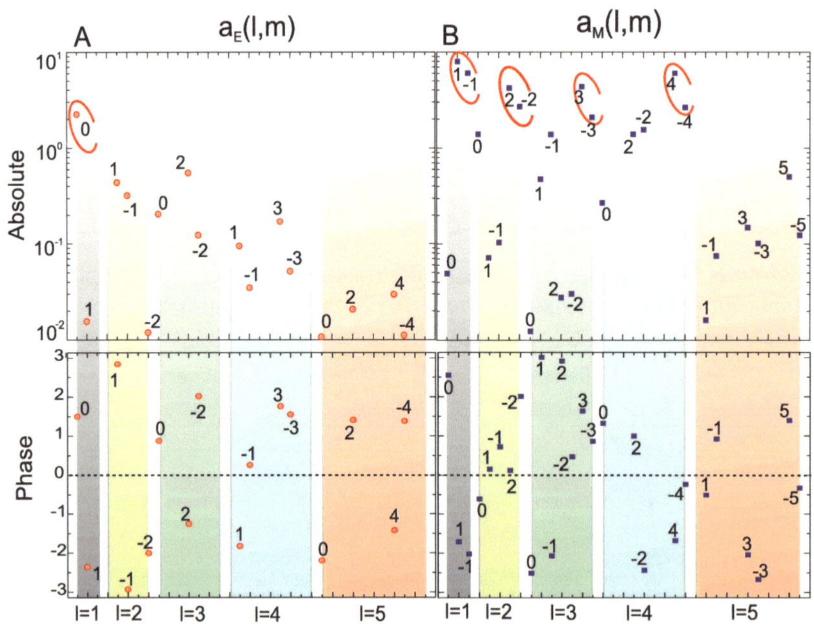

Figure 20. Absolute values and phases of (**A**) electric and (**B**) magnetic multipole moments that provide the main contribution to the radiation of all-dielectric superdirective optical nanoantenna in case of asymmetrical location of source at the wavelength 455 nm. Coefficients that give the largest contribution to the antenna directivity are highlighted by red circles.

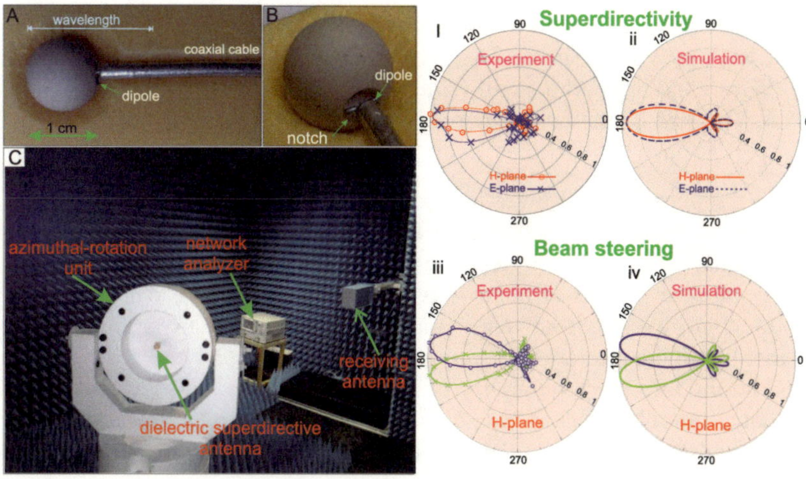

Figure 21. Photographs of (**A**) top view and (**B**) perspective view of a notched all-dielectric microwave antenna. Image of (**C**) the experimental setup for measuring of power patterns. Experimental (i) and numerical (ii) radiation patterns of the antenna in both E- and H-planes at the frequency 16.8 GHz. The crosses and circles correspond to the experimental data. Experimental (iii) and numerical (iv) demonstration of beam steering effect, displacement of dipole is equal 0.5 mm.

To interpret the beam steering effect, we can consider the result of field expansion to electric and magnetic multipoles, as shown in Fig.20. In the case of asymmetrical location (the 20 nm left offset) of the source in the notch absolute values of $a_M(l, \pm|m|)$ are different. This means that the mode $a_M(l, +|m|)$ is excited more strongly than $a_M(l, -|m|)$, or vice versa, that depends on direction of displacement. The effect of superdirectivity remains even with an offset of the source until to the edge of the notch. Small displacements of the source along x and z do not lead to the rotation of the pattern.

Instead of the movement of a single quantum dot one we can have the emission of two or more quantum dots located near the edges of the notch. In this case, the dynamics of their spontaneous decay will be well displayed in the angular distribution of the radiation. This can be useful for quantum information processing and for biomedical applications.

Beam steering effect described above is similar to the effect of beam rotation in hyperlens [117–119], where the displacement of a point-like source leads to a change of the angular distribution of the radiation power. However, in our case, the nanoantenna has subwavelength dimensions and therefore it can be neither classified as a hyperlens nor as a micro-spherical dielectric nanoscope [104, 105], moreover it is not an analogue of solid immersion micro-lenses [108–111], which are characterized by the size 1-5 μm in the same frequency range. These lens has a subwavelength resolving power due to the large geometric aperture but the value of normalized effective aperture is $S_n \simeq 1$. Our study demonstrates that the sub-wavelength system, with *small compared to the wavelength* geometric aperture can have both high directing and resolving power *because of a strong increase of the effective aperture compared to the geometrical one.*

4.3. Experimental verification of superdirective optical nanoantenna

We have confirmed both predicted effects studying the similar problem for the microwave range. Therefore, we have scaled up the nanoantenna as above to low frequencies. Instead of Si we employ $MgO-TiO_2$ ceramic [46] characterized at microwaves by a dispersion-less dielectric constant 16 and dielectric loss factor of $1.12 \cdot 10^{-4}$. We have used the sphere of radius $R_S = 5$ mm and applied a small wire dipole [80] excited by a coaxial cable as shown in Fig. 21A,B. The size of the hemispherical notch is approximately equal to $R_n = 2$ mm. Antenna properties have been studied in an anechoic chamber Fig. 21C.

The results of the experimental investigations and numerical simulations of the pattern in both *E*- and *H*-planes are summarized in Figs. 21i,ii. Radiation patterns in both planes are narrow beams with a lobe angle about 35°. Experimentally obtained coefficients of the directivity in both *E*- and *H*-planes are equal to 5.9 and 8.4, respectively (theoretical predictions for them were respectively equal 6.8 and 8.1). Our experimental data are in a good agreement with the numerical results except a small difference for the E plane, that can be explained by the imperfect symmetry of the emitter. Note, that the observed directivity is close to that of an all-dielectric Yagi-Uda antenna with maximum size of 2λ [46]. The maximum size of our experimental antenna is closed to $\lambda/2.5$. Thus, our experiment clearly demonstrates the superdirective effect.

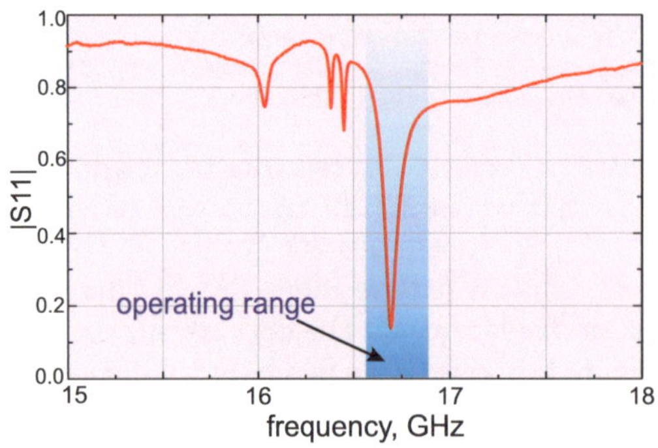

Figure 22. Level of the return losses of superdirective dielectric antenna. Blue area shows the operating frequency range.

Experimental and numerical demonstration of the beam steering effect are presented in Figs. 21iii,iv. For the chosen geometry of antenna, displacement of source by 0.5 mm leads to a beam rotation of about 10°. Note that the ratio of $\lambda = 18.7$ mm to value of the source displacement 0.5 mm is equal to 37. Therefore the beam steering effect observed at subwavelength source displacement.

Finally, we consider the question of dielectric superdirective antenna matching with coaxial cable. Despite that length of the wire dipole is close to $\lambda/10$, dielectric superdirective antenna is well matched with the coaxial cable in the operating frequency range. Fig.22 shows the level of return loss for this case. The antenna matching is explained by the strong coupling of the wire dipole with the excited modes of notched dielectric particle and is not related to the dissipative losses in the superdirectivity regime. For this reason, we have not used additional matching devices (e.g. "balun").

Though the concept of the superdirectivity of high-refractive index dielectric particles with notch has now only been proven in GHz spectral range there is a hope that it can be transferred into the visible and near-IR spectrum in the nearest future. Recently we have experimentally demonstrated that it is possible to engineer resonant modes of spherical nanoresonators using a combined approach of laser-induced transfer to generate almost perfect spherical nanoparticles and helium ion beam milling to structure their surface with sub-5nm resolution [120]. This novel approach can become a suitable candidate for realizing all-dielectric superdirective nanoantennas.

Conclusion

In this chapter, we propose a new type of highly efficient Yagi-Uda nanoantenna and introduced a novel concept of superdirective nanoantennas based on silicon nanoparticles. In addition to the electric response, this silicon nanoantennas exhibit very strong magnetic resonances at the nanoscale. Both types of nanoantennas are studied analytically, numerically and experimentally. For superdirective nanoantennas we also predict the effect of the beam steering at the nanoscale characterized by a subwavelength sensitivity of the beam radiation direction to the source position.

Dielectric nanoparticles with high refractive index offer new possibilities for achieving wave interference. Indeed, the coexistence of both electric and magnetic resonances results in a unidirectional scattering. This property makes subwavelength dielectric nanoparticles the smallest and most efficient nanoantennas. Moreover, unidirectionality can be swapped for different wavelengths [74, 76].

The unique optical properties and low losses make dielectric nanoparticles perfect candidates for a design of high-performance nanoantennas, low-loss metamaterials, and other novel all-dielectric nanophotonic devices. The key to such novel functionalities of high-index dielectric nanophotonic elements is the ability of subwavelength dielectric nanoparticles to support simultaneously both electric and magnetic resonances, which can be controlled independently for particles of non spherical forms [77, 121].

Author details

Alexandr E. Krasnok[1], Pavel A. Belov[1], Andrey E. Miroshnichenko[2], Arseniy I. Kuznetsov[3], Boris S. Luk'yanchuk[3] and Yuri S. Kivshar[1,2]

1 ITMO University, St. Petersburg, Russia
2 Nonlinear Physics Centre, Research School of Physics and Engineering, Australian National University, Canberra, Australian Capital Territory, Australia
3 Data Storage Institute, A*STAR (Agency for Science, Technology and Research), Singapore

References

[1] Palash Bharadwaj, Bradley Deutsch, and Lukas Novotny. Optical antennas. *Advances in Optics and Photonics*, 1:438–483, 2009.

[2] Lukas Novotny and Niek van Hulst. Antennas for light. *Nat. Photon.*, 5:83–90, 2011.

[3] Alberto G. Curto, Giorgio Volpe, Tim H. Taminiau, Mark P. Kreuzer, Romain Quidant, and Niek F. van Hulst. Unidirectional emission of a quantum dot coupled to a nanoantenna. *Science*, 329:930–933, 2010.

[4] T.H. Taminiau, F.D. Stefani, F.B. Segerink, and N.F. Van Hulst. Optical antennas direct single-molecule emission. *Nat. Photonics.*, 2:234–237, 2008.

[5] Lukas Novotny. Optical antennas tuned to pitch. *Nature*, 455:887, 2008.

[6] S. He, Y. Cui, Y. Ye, P. Zhang, and Y. Jin. Optical nanoantennas and metamaterials. *Materials Today*, 12:16, 2009.

[7] L. Novotny and B. Hecht. *Principles of Nano-Optics*. Cambridge University Press, 2006.

[8] A.E. Krasnok, I.S. Maksymov, A.I. Denisyuk, P.A. Belov, A.E. Miroshnichenko, C.R. Simovski, and Yu.S. Kivshar. Optical nanoantennas. *Phys.-Usp.*, 56:539, 2013.

[9] A.E. Krasnok and P.A. Belov. *Optical nanonatennas*. LAP LAMBERT Academic Publishing, 2013.

[10] Mark W. Knight, Heidar Sobhani, Peter Nordlander, and Naomi J. Halas. Photodetection with active optical antennas. *Science*, 332:702–704, 2011.

[11] M. F. Garcia-Parajo. Optical antennas focus in on biology. *Nature Photonics*, 2:201–203, 2008.

[12] A. Alu and N. Engheta. Wireless at the nanoscale: Optical interconnects using matched nanoantennas. *Phys. Rev. Lett.*, 104:213902, 2010.

[13] Brian J. Roxworthy, Kaspar D. Ko, Anil Kumar, Kin Hung Fung, Edmond K. C. Chow, Gang Logan Liu, Nicholas X. Fang, and Kimani C. Toussaint. Application of plasmonic bowtie nanoantenna arrays for optical trapping, stacking, and sorting. *Nano Lett.*, 12:796–801, 2012.

[14] A. Andryieuski, R. Malureanu, G. Biagi, T. Holmgaard, and A. Lavrinenko. Compact dipole nanoantenna coupler to plasmonic slot waveguide. *Opt. Lett.*, 37:1124–1126, 2012.

[15] B. Khlebtsov, A. Melnikov, V. Zharov, and N. Khlebtsov. Absorption and scattering of light by a dimer of metal nanospheres: comparison of dipole and multipole approaches. *Nanotechnology*, 17:1437, 2006.

[16] F. Gonzalez and G.D. Boreman. Comparison of dipole, bowtie, spiral and log-periodic ir antennas. *Infrared Phys. Tech.*, 46:418–428, 2005.

[17] Zheyu Fang, Linran Fan, Chenfang Lin, Dai Zhang, Alfred J. Meixner, and Xing Zhu. Plasmonic coupling of bow tie antennas with ag nanowire. *Nano Lett.*, 11, 2011.

[18] Hongcang Guo, Todd P. Meyrath, Thomas Zentgraf, Na Liu, Liwei Fu, Heinz Schweizer, and Harald Giessen. Optical resonances of bowtie slot antennas and their geometry and material dependence. *Opt. Express*, 16:7756–7766, 2008.

[19] Nahla A. Hatab, Chun-Hway Hsueh, Abigail L. Gaddis, Scott T. Retterer, Jia-Han Li, Gyula Eres, Zhenyu Zhang, and Baohua Gu. Free-standing optical gold bowtie

nanoantenna with variable gap size for enhanced raman spectroscopy. *Nano Lett.*, 10, 2010.

[20] Anika Kinkhabwala, Zongfu Yu, Shanhui Fan, Yuri Avlasevich, Klaus Mullen, and W. E. Moerner. Large single-molecule fluorescence enhancements produced by a bowtie nanoantenna. *Nat. Phot.*, 3:654–657, 2009.

[21] Kaspar D. Ko, Anil Kumar, Kin Hung Fung, Raghu Ambekar, Gang Logan Liu, Nicholas X. Fang, and Kimani C. Toussaint. Nonlinear optical response from arrays of au bowtie nanoantennas. *Nano Lett.*, 11, 2011.

[22] S. Sederberg and A. Y. Elezzabi. Nanoscale plasmonic contour bowtie antenna operating in the mid-infrared. *Opt. Express*, 19:15532–15537, 2011.

[23] Jae Yong Suh, Mark D. Huntington, Chul Hoon Kim, Wei Zhou, Michael R. Wasielewski, and Teri W. Odom. Extraordinary nonlinear absorption in 3d bowtie nanoantennas. *Nano Lett.*, 12, 2012.

[24] Z. Zhang, A. Weber-Bargioni, S. W. Wu, S. Dhuey, S. Cabrini, and P. J. Schuck. Manipulating nanoscale light fields with the asymmetric bowtie nano-colorsorter. *Nano Lett.*, 9:4505–4509, 2009.

[25] Andrea Alu and Nader Engheta. Input impedance, nanocircuit loading, and radiation tuning of optical nanoantennas. *PRL*, 101:043901, 2008.

[26] Andrea Alu and Nader Engheta. Tuning the scattering response of optical nanoantennas with nanocircuit loads. *Nat. Phot.*, 2:307–310, 2008.

[27] Toon Coenen, Ernst Jan R. Vesseur, Albert Polman, and A. Femius Koenderink. Directional emission from plasmonic yagi-uda antennas probed by angle-resolved cathodoluminescence spectroscopy. *Nano Lett.*, 11:3779–3784, 2011.

[28] Jens Dorfmuller, Daniel Dregely, Moritz Esslinger, Worawut Khunsin, Ralf Vogelgesang, Klaus Kern, and Harald Giessen. Near-field dynamics of optical yagi-uda nanoantennas. *Nano Lett.*, 11:2819–2824, 2011.

[29] Holger F Hofmann, Terukazu Kosako, and Yutaka Kadoya. *New Journal of Physics*, 9:217, 2007.

[30] G. Lerosey. Nano-optics: Yagi?uda antenna shines bright. *Nat. Phot.*, 4:267–268, 2010.

[31] Jingjing Li, Alessandro Salandrino, and Nader Engheta. Optical spectrometer at the nanoscale using optical yagi-uda nanoantennas. *Phys. Rev. B*, 79:195104, 2009.

[32] S. V. Lobanov, T. Weiss, D. Dregely, H. Giessen, N. A. Gippius, and S. G. Tikhodeev. Emission properties of an oscillating point dipole from a gold yagi-uda nanoantenna array. *Phys. Rev. B*, 85:155137, 2012.

[33] I. S. Maksymov, I. Staude, A. E. Miroshnichenko, and Y. S. Kivshar. Optical yagi-uda nanoantennas. *Nanophotonics*, 1:65–81, 2012.

[34] Ivan S. Maksymov, Andrey E. Miroshnichenko, and Yuri S. Kivshar. Actively tunable bistable optical yagi-uda nanoantenna. *Opics Express*, 20:8929–8938, 2012.

[35] Tim H. Taminiau, Fernando D. Stefani, and Niek F. van Hulst. Enhanced directional excitation and emission of single emitters by a nano-optical yagi-uda antenna. *Opt. Express*, 16:16858–16866, 2008.

[36] Alexander E. Krasnok, Constantin R. Simovski, Pavel A. Belov, and Yuri S. Kivshar. Superdirective dielectric nanoantenna. *Nanoscale*, 6:7354–7361, 2014.

[37] A. E. Krasnok, A. E. Miroshnichenko, P. A. Belov, and Yu. S. Kivshar. All-dielectric optical nanoantennas. *Optics Express*, 20:20599, 2012.

[38] V.P. Dzyuba, A.E. Krasnok, and Yu.N. Kulchin. A model of nonlinear optical transmittance for insulator nanocomposites. *Semiconductors*, 45:295–301, 2011.

[39] V. P. Dzyuba, A. E. Krasnok, and Yu. N. Kul'chin. Nonlinear refractive index of dielectric nanocomposites in weak optical fields. *Technical Physics Letters*, 36:973–977, 2010.

[40] Alexander E. Krasnok, Andrey E. Miroshnichenko, Pavel A. Belov, and Yuri S. Kivshar. All-dielectric optical nanoantennas. *AIP Conf. Proc.*, 1475:22–24, 2012.

[41] Alexander E. Krasnok, Andrey E. Miroshnichenko, Pavel A. Belov, and Yuri S. Kivshar. All-dielectric nanoantennas. *Proc. of SPIE*, 8806:880626, 2013.

[42] R. E. Noskov, A. E. Krasnok, and Yu. S. Kivshar. Nonlinear metal-dielectric nanoantennas for light switching and routing. *New Journal of Physics*, 14:093005, 2012.

[43] Dmitry S. Filonov, Alexey P. Slobozhanyuk, Alexander E. Krasnok, Pavel A. Belov, Elizaveta A. Nenasheva, Ben Hopkins, Andrey E. Miroshnichenko, and Yuri S. Kivshar. Near-field mapping of fano resonances in all-dielectric oligomers. *Appl. Phys. Lett.*, 104:021104, 2014.

[44] A. E. Krasnok, A. E. Miroshnichenko, P. A. Belov, and Yu. S. Kivshar. Huygens optical elements and yagi?uda nanoantennas based on dielectric nanoparticles. *JETP Lett.*, 94:635, 2011.

[45] Alexander E. Krasnok, Andrey E. Miroshnichenko, Pavel A. Belov, and Yuri S. Kivshar. All-dielectric optical nanoantennas. *Opt. Express*, 20:20599–20604, 2012.

[46] Dmitry S. Filonov, Alexander E. Krasnok, Alexey P. Slobozhanyuk, Polina V. Kapitanova, Elizaveta A. Nenasheva, Yuri S. Kivshar, and Pavel A. Belov. Experimental verification of the concept of all-dielectric nanoantennas. *Appl. Phys. Lett.*, 100:201113, 2012.

[47] A.I. Kuznetsov, A.E. Miroshnichenko, Y.H. Fu, J. Zhang, and B. Lukyanchuk. Magnetic light. *Sci. Rep.*, 2:492, 2012.

[48] L. D. Landau, E. M. Lifshitz, and L. P. Pitaevskii. *Electrodynamics of Continuous Media.* Butterworth-Heinemann, Oxford, 2002.

[49] V. M. Shalaev. Optical negative-index metamaterials. *Nature Photon.*, 1:41–47, 2007.

[50] N. I. Zheludev. The road ahead for metamaterials. *Science*, 328:582–583, 2010.

[51] C. M. Soukoulis and M. Wegener. *Science*, 330:1633–1634, 2010.

[52] A. Boltasseva and H. A. Atwater. Low-loss plasmonic metamaterials. *Science*, 331:290–291, 2011.

[53] C. M. Soukoulis and M. Wegener. Past achievements and future challenges in the development of three-dimensional photonic metamaterials. *Nature Photon.*, 5:523–530, 2011.

[54] J. B. Pendry, A. J. Holden, D. J. Robbins, and W. J. Stewart. Magnetism from conductors and enhanced nonlinear phenomena. *IEEE Trans. Microwave Theory Tech.*, 47:2075–2084, 1999.

[55] D. R. Smith, W. J. Padilla, D. C. Vier, S. C. Nemat-Nasser, and S. Schultz. Composite medium with simultaneously negative permeability and permittivity. *Phys. Rev. Lett.*, 84:4184–4187, 2000.

[56] R. A. Shelby, D. R. Smith, and S. Schultz. Experimental verification of a negative index of refraction. *Science*, 292, 2001.

[57] D. R. Smith, S. Schultz, P. Markos, and C. M. Soukoulis. Determination of effective permittivity and permeability of metamaterials from reflection and transmission coefficients. *Phys. Rev. B*, 65:195104, 2002.

[58] U. Leonhardt. Optical conformal mapping. *Science*, 312:1777–1780, 2006.

[59] J. B. Pendry, D. Schurig, and D. R. Smith. Controlling electromagnetic fields. *Science*, 312:1780–1782, 2006.

[60] J. B. Pendry. Negative refraction makes a perfect lens. *Phys. Rev. Lett.*, 85:3966–3969, 2000.

[61] T. J. Yen, T. J. Padilla, N. Fang, D. C. Vier, D. R. Smith, J. B. Pendry, D. N. Basov, and X. Zhang. Terahertz magnetic response from artificial materials. *Science*, 303, 2004.

[62] W. J. Padilla, A. J. Taylor, C. Highstrete, M. Lee, and R. D. Averitt. Dynamical electric and magnetic metamaterial response at terahertz frequencies. *Phys. Rev. Lett.*, 96:107401, 2006.

[63] S. Linden, C. Enkrich, M. Wegener, J. Zhou, T. Koschny, and C. M. Soukoulis. Magnetic response of metamaterials at 100 terahertz. *Science*, 306, 2004.

[64] C. Enkrich, M. Wegener, S. Linden, S. Burger, L. Zschiedrich, F. Schmidt, J. F. Zhou, Th. Koschny, and C. M. Soukoulis. Magnetic metamaterials at telecommunication and visible frequencies. *Phys. Rev. Lett.*, 95:203901, 2005.

[65] N. Liu, H. Guo, L. Fu, S. Kaiser, H. Schweizer, and H. Giessen. Three-dimensional photonic metamaterials at optical frequencies. *Nature Mater.*, 7, 2008.

[66] C. M. Soukoulis, T. Koschny, J. Zhou, M. Kafesaki, and E. N. Economou. Magnetic response of split ring resonators at terahertz frequencies. *Phys. Stat. Sol. B*, 244:1181–1187, 2007.

[67] Q. Zhao, J. Zhou, F. Zhang, and D. Lippens. Mie resonance-based dielectric metamaterials. *Mater. Today*, 12, 2009.

[68] L. Cao, P. Fan, E. S. Barnard, A. M. Brown, and M. L. Brongersma. Tuning the color of silicon nanostructures. *Nano Lett.*, 10:2649–2654, 2010.

[69] Optical response features of si-nanoparticle arrays. *Phys. Rev. B*, 82:045404, 2010.

[70] A. Garcia-Etxarri, R. Gomez-Medina, L. S. Froufe-Perez, C. Lopez, L. Chantada, F. Scheffold, J. Aizpurua, M. Nieto-Vesperinas, and J. J. Saenz. Strong magnetic response of submicron silicon particles in the infrared. *Opt. Express*, 19:4815–4826, 2011.

[71] A. V. Kabashin and M. Meunier. Synthesis of colloidal nanoparticles during femtosecond laser ablation of gold in water. *J. Appl. Phys.*, 94, 2003.

[72] C.F. Bohren and D.R. Huffman. *Absorption and scattering of light by small particles*. New York : Wiley, 1998.

[73] A. B. Evlyukhin, S. M. Novikov, U. Zywietz, R. L. Eriksen, C. Reinhardt, S. I. Bozhevolnyi, and B. N. Chichkov. Demonstration of magnetic dipole resonances of dielectric nanospheres in the visible region. *Nano Lett.*, 12:3749, 2012.

[74] Yuan Hsing Fu, Arseniy I. Kuznetsov, Andrey E. Miroshnichenko, Ye Feng Yu, and Boris Lukyanchuk. Directional visible light scattering by silicon nanoparticles. *Nat. Comm.*, 4:1527, 2013.

[75] M. Kerker, D.-S. Wang, and C. L. Giles. Electromagnetic scattering by magnetic spheres. *J. Opt. Soc. Am.*, 73(6):765–767, Jun 1983.

[76] J.M. Geffrin. Magnetic and electric coherence in forward- and back-scattered electromagnetic waves by a single dielectric subwavelength sphere. *Nat. Comm.*, 3:1, 2012.

[77] Steven Person, Manish Jain, Zachary Lapin, Juan Jose Saenz, Gary Wicks, and Lukas Novotny. Demonstration of zero optical backscattering from single nanoparticles. *Nano Letters*, 13(4):1806–1809, 2013.

[78] M. K. Schmidt, R. Esteban, J. J. S?enz, I. Su?rez-Lacalle, S. Mackowski, and J. Aizpurua. Dielectric antennas - a suitable platform for controlling magnetic dipolar emission. *Optics Express*, 20:13636–13650, 2012.

[79] E. Palik. *Handbook of Optical Constant of Solids*. San Diego, Academic, 1985.

[80] C. Balanis. *Antenna theory : analysis and design*. New York ; Brisbane : J. Wiley, 1982.

[81] Optically induced interaction of magnetic moments in hybrid metamaterials. *ASC Nano*, 6:837–842, 2012.

[82] O. Merchiers, F. Moreno, F. Gonzalez, and J. M. Saiz. Light scattering by an ensemble of interacting dipolar particles with both electric and magnetic polarizabilities. *Phys. Rev. A*, 76:043834, 2007.

[83] Brice Rolly, Brian Stout, Sebastien Bidault, and Nicolas Bonod. *Opt. Lett.*, 36:3368–3370, 2011.

[84] L. Novotny and N. van Hulst. Antennas for light. *Nat. Photon.*, 5:83–90, 2011.

[85] H. Chew. Transition rates of atoms near spherical surfaces. *J. Chern. Phys.*, 87:1355, 1987.

[86] Yu lin Xu and Bo A. S. Gustafson. A generalized multiparticle mie-solution: further experimental verification. *J. Quant. Spectrosc. Radiat. Transfer*, 70:395–419, 2001.

[87] E. Dulkeith, M. Ringler, T. A. Klar, J. Feldmann, A. M. Javier, and W. J. Parak. Gold nanoparticles quench fluorescence by phase induced radiative rate suppression. *Nano Lett.*, 5:585–589, 2005.

[88] A. Kanareykin, W. Gai, J. G. Power, E. Nenasheva, and A. Altmark. Transformer ratio enhancement experiment. In *Proceedings Particle Accelerator Conference*, pages 1894–1896, 2003.

[89] N.J. Halas. Connecting the dots: Reinventing optics for nanoscale dimensions. *PNAS*, 106:3643, 2010.

[90] P. Biagioni, J.S. Huang, and B. Hecht. Nanoantennas for visible and infrared radiation. *Rep. Prog. Phys.*, 75:024402, 2012.

[91] D.M. Solis, J.M. Taboada, F. Obelleiro, and L. Landesa. Optimization of an optical wireless nanolink using directive nanoantennas. *Optics Express*, 21:2369, 2013.

[92] D. Dregely, R. Taubert, J. Dorfmuller, R. Vogelgesang, K. Kern, and H. Giessen. 3d optical yagi−uda nanoantenna array. *Nat. Comm.*, 2:1–7, 2011.

[93] Y. G. Liu, W. C. H. Choy, W. E. I. Sha, and W. C. Chew. Unidirectional and wavelength-selective photonic sphere-array nanoantennas. *Optics Lett.*, 37:2112, 2012.

[94] Alexis Devilez, Brian Stout, and Nicolas Bonod. Compact metallo-dielectric optical antenna for ultra directional and enhanced radiative emission. *ACS Nano*, 4:3390–3396, 2010.

[95] A. Ludwig, C.D. Sarris, and G.V. Eleftheriades. Metascreen-based superdirective antenna in the optical frequency regime. *Phys. Rev. Lett.*, 109:223901, 2012.

[96] P. K. Jain and M. A. El-Sayed. Plasmonic coupling in noble metal nanostructures. *Chemical Physics Letters*, 487, 2010.

[97] S. R. K. Rodriguez, S. Murai, M. A. Verschuuren, and J. Gomez Rivas. Light-emitting waveguide-plasmon polaritons. *Phys. Rev. Lett.*, 109:166803, 2012.

[98] B. Rolly, B. Stout, and N. Bonod. Boosting the directivity of optical antennas with magnetic and electric dipolar resonant particles. *Optics Exp.*, 20:20376, 2012.

[99] R.C. Hansen. *Electrically small, superdirective, and superconducting antennas.* Wiley-Interscience, 2006.

[100] R. Hansen and R. Collin. *Small Antenna Handbook.* John Wiley and Sons Ltd, 2011.

[101] S.V. Boriskina, T.M. Benson, P. Sewell, and A.I. Nosich. Directional emission, increased free spectral range and mode q-factors in 2-d wavelength-scale optical microcavity structures. *IEEE J. Select. Topics Quantum Electron.*, 12:1175–1182, 2006.

[102] Q. J. Wang, C. Yan, N. Yu, J. Unterhinninghofen, J. Wiersig, C. Pflugl, L. Diehl, T. Edamura, M. Yamanishi, H. Kan, and F. Capasso. Whispering-gallery mode resonators for highly unidirectional laser action. *PNAS*, 107:22407, 2010.

[103] M. O. Scully. Collimated unidirectional laser beams from notched elliptical resonators. *PNAS*, 107:22367, 2010.

[104] J. Y. Lee, B. H. Hong, W. Y. Kim, S. K. Min, Y. Kim, M. V. Jouravlev, R. Bose, K. S. Kim, I. Hwang, L. J. Kaufman, C. W. Wong, P. Kim, and K. S. Kim. Near-field focusing and magnification through self-assembled nanoscale spherical lenses. *Nature*, 460:498, 2009.

[105] Z. Wang, W. Guo, L. Li, B. Luk'yanchuk, A. Khan, Z. Liu, Z. Chen, and M. Hong. Optical virtual imaging at 50 nm lateral resolution with a white-light nanoscope. *Nat. Comm.*, 2:1, 2011.

[106] L.H. Gabrielli and M. Lipson. Integrated luneburg lens via ultra-strong index gradient on silicon. *Opt. Expr.*, 19:20122, 2011.

[107] A.D. Falco, S.C. Kehr, and U. Leonhardt. Luneburg lens in silicon photonics. *Opt. Expr.*, 19:5156, 2011.

[108] D. Gerard, J. Wenger, A. Devilez, D. Gachet, B. Stout, N. Bonod, E. Popov, and H. Rigneault. Strong electromagnetic confinement near dielectric microspheres to enhance single-molecule fluorescence. *Opt. Expr.*, 16:15297, 2008.

[109] J.J. Schwartz, S. Stavrakis, and S.R. Quake. Colloidal lenses allow high-temperature single-molecule imaging and improve fluorophore photostability. *Nat Nanotechnol.*, 5:127, 2010.

[110] L. Robledo, L. Childress, H. Bernien, B. Hensen, P. F. A. Alkemade, and R. Hanson. High-fidelity projective read-out of a solid-state spin quantum register. *Nature*, 477:574, 2011.

[111] P. Siyushev, F. Kaiser, V. Jacques, I. Gerhardt, S. Bischof, H. Fedder, J. Dodson, M. Markham, D. Twitchen, F. Jelezko, and J. Wrachtrup. Monolithic diamond optics for single photon detection. *Appl. Phys. Lett.*, 97:241902, 2010.

[112] V.V. Klimov. *Nanoplasmonics*. Pan Stanford Publishing, 2011.

[113] T. Pakizeh and M. Kall. Unidirectional ultracompact optical nanoantennas. *Nano Lett.*, 9:2343, 2009.

[114] S.H. Alavi Lavasani and T. Pakizeh. Color-switched directional ultracompact optical nanoantennas. *Opt. Lett.*, 29:1361, 2012.

[115] A. Alu and N. Engheta. Enhanced directivity from subwavelength infrared/optical nano-antennas loaded with plasmonic materials or metamaterials. *IEEE Trans. on Antennas and Propagation*, 55:3027, 2007.

[116] J.D. Jackson. *Classical Electrodynamics*. New York : Wiley, 1998.

[117] Z. Liu, H. Lee, Y. Xiong, C. Sun, and X. Zhang. Far-field optical hyperlens magnifying sub-diffraction-limited objects. *Science*, 315:1686, 2007.

[118] Z. Jacob, L.V. Alekseyev, and E. Narimanov. Optical hyperlens: Far-field imaging beyond the diffraction limit. *Opt. Expr.*, 14:8247, 2006.

[119] D. Lu and Z. Liu. Hyperlenses and metalenses for far-field super-resolution imaging. *Nat. Comm.*, 3:1, 2012.

[120] Arseniy I. Kuznetsov, Andrey E. Miroshnichenko, Yuan Hsing Fu, Vignesh Viswanathan, Mohsen Rahmani, Vytautas Valuckas, Zhen Ying Pan, Yuri Kivshar, Daniel S. Pickard, and Boris Luk?yanchuk. Split-ball resonator as a three-dimensional analogue of planar split-rings. *Nature Communications*, 5:3104, 2013.

[121] Isabelle Staude, Andrey E. Miroshnichenko, Manuel Decker, Nche T. Fofang, Sheng Liu, Edward Gonzales, Jason Dominguez, Ting Shan Luk, Dragomir N. Neshev, Igal Brener, and Yuri Kivshar. Tailoring directional scattering through magnetic and electric resonances in subwavelength silicon nanodisks. *ACS Nano*, 7(9):7824?7832, 2013.

Chapter 7

Building Partial Discharge Signal Wireless Probes

Fabio Viola and Pietro Romano

Additional information is available at the end of the chapter

http://dx.doi.org/10.5772/58840

1. Introduction

In the recent years compact ultra-wideband (UWB) antennas have received a great attention [1-3] due to the allocation of various frequency bands to UWB systems by Federal Communications Committee (FCC) [4]. One of the most interesting aspects of these antennas is their aptitude of detecting electromagnetic (EM) transients with frequency content up to the very high frequency (VHF) range. EM transient phenomena with the above mentioned frequency content are those encountered, for example, in partial discharge (PD) detection. This diagnostic method is now widely used to identify defects taking place in the insulation systems. Wireless systems offer the possibility to achieve the whole shape of PD signals radiated from the source of the PD defects with less distortion due to the electric transmission path.

Modern diagnostic procedures require clear PD patterns for the identification of defects generating PD because different defects can affect differently the insulation reliability, [5-7]. Different pulsating sources can be simultaneously active during a PD measurement session and mixed PD patterns can be recorded. Thus, an effective separation of mixed PD patterns into sub-patterns each one pertinent to a specific noise or PD source typology, is a fundamental task to avoid wrong defect identification, [8-9]. One of the possible approaches in signal separation is based on the assumption that the same defect generates similar waveforms and features derived from signals grouped by similarity can be adopted to identify the defect or noise source. Thus, the dynamic characteristics of the antenna probe must be designed/evaluated in order to fit the above mentioned requirements.

Both the time and the frequency domain can be used for separation purposes. Recently, the use of the Auto-Correlation Function (ACF) has been proposed for its ability to synthesize both the time and frequency domain features as well as to be less affected by superimposed high-frequency noise, random truncation of the pulse-tail (frequency leakage) and different trigger activation, [10].

The ACF has been adopted here to evaluate the performances of different EM probes in order to design the more suitable geometry to build a portable low-cost device for PD detection. Monopole, triangular and spherical antennas were simulated by means of the surface Method of Moments (MoM) in the frequency domain, [11-14]. The transmitting system is modeled by a power electronic device with a fault current between two metal plates. The ACF shapes of the simulated, transmitted and received signals, have been compared using the Pearson correlation index to verify which sensor is providing the best fidelity among the three.

This chapter focuses on the evaluation of the performances of different antenna sensors suitable for Partial Discharge (PD) measurements. Monopole, triangular and spherical antennas were simulated by means of the surface method of moments. The transmitting system is modeled by a power electronic device with a fault current between two metal plates. The shape of the simulated, transmitted and received signals, has been compared to verify the sensor that provides the best fidelity among the three. The auto-correlation function and the Pearson correlation index are adopted here for the comparison. A discussion on the dynamic characteristic of the different antenna probes and their use in different application is proposed.

2. Partial discharges

The increasing necessity of high voltage levels in high voltage (HV) power systems has driven to a stronger demand of insulating materials with high electrical performances at affordable costs. Since synthetic polymers partially meet these requirements [15, 16], they have been studied quite a bit in the most recent literature. However, they are affected by some aging problems also the PD phenomenon is a major cause of failure in the insulation systems of electrical apparatus. These discharges take place in consequence of unavoidable local defects produced by the industrial manufacturing process, and promote local erosion of the material that may cause electric breakdown of the component over time. The difficulty of performing reliable life predictions and suitable evaluations of their reliability in presence of such degradation phenomena, hindered a wide diffusion of HV components made of epoxy materials. The diagnostic techniques of recognition, identification and classification of discharge phenomena are therefore fundamental for the evaluation of the time reliability of the electrical equipment, especially for high voltages. In order to study and understand the behaviour of different materials, the first step is to perfectly know the dielectric characteristics.Materials are divided into three basic categories that are metals, ceramics and polymers. Furthermore, the combination of two or more of these elements gives rise to materials employed in common applications normally named composite materials. Successively, another division is based on the type of atomic bond and on the basis of the crystal structure that composes them. Besides, in solid state physics, the electronic band structure of a solid describes the range of energies that an electron of a certain material is "allowed" or "forbidden" to possess. The electronic structure of a material greatly is influencing the materials behaviour and the energy gap between the valence band. The conduction is used to classify the materials on the electronic characteristics, in conductors, semiconductors and insulators. The insulators materials or dielectric materials present a wide forbidden zone between the valence band and the conduction band. Dielectric materials are divided into three categories named solid, liquid and gaseous. The main electrical properties that characterize the behaviour of insulating materials are:

- *Intrinsic/extrinsic conductivity*σ and γ are about $10^{-5}\div10^{-18}$ [S/m].
- *Electric permittivity* $\varepsilon_r=\varepsilon/\varepsilon_0$; where ε is the permittivity of a generic material and ε_0 is the vacuum dielectric permittivity with value $8 \bullet 85418781762 \times 10^{-12}$ [F/m].
- *Loss factor*, defined as tgδ with δ=90°-φ, the loss angle of the material, with φ the phase shift between voltage and current, which, in the ideal case is equal to 90 ° while, in a real dielectric, is less than 90 ° due to their leaks.
- *Dielectric Rigidity* (RD):value of the voltage gradient required to cause discharge through the dielectric under examination or the limit value of the electric field beyond which produces a conduction of electricity (electrical discharge) through the dielectric material. It is commonly measured in kV/mm. The discharge gradient is strongly influenced bya very high number of different parameters and by the test procedure, in particular by the shape of electrodes, by the voltage applied and by the way the limit voltage is reached.

The Test procedures for these characteristic parameters depending on the dielectric materials state are often very different. To better understand the partial discharge phenomenon, the discharge base theory in gas is synthetically presented in the next section, while extended information are reported in [15, 16]. Gaseous insulation, especially ambient air, is the most commonly used insulation in high voltage engineering.

2.1. Discharge theory

The occurrence of partial discharges in electrical equipment has been identified as one of the main causes of breaking devices and reduction in useful life of insulation. For many years, studies on the physical and chemical aspects of PD have been thoroughly addressed and news discharge models were introduced. The principle of discharge starting mechanism, for small distances, was introduced by Towsend [15, 16] and it is based on the coefficient of ionization of the gas subjected to an electric field. For distances more than a few millimetres, Towsend theory is wrong, since discharge develops in a times lower than necessary ones for electrons to cross the distance between electrodes and form an avalanche. To explain the formation of the discharge's mechanism over long distances, several assumptions were made based on phenomenon such as the formation of individual electronic avalanches or photo-ionization phenomena. Meek and Loeb, and independently Rather, proposed the Streamer theory. The equations that describe phenomena involving electrons and ions, avalanches and streamers, are too complex to be dealt with in a few lines and can be found in [15,16].

It is possible to macroscopically represent the discharge phenomenon with a circuit model. This one is representing the system cavity-electrodes in the dielectric. The Whitehead model is exhibiting three capacitors to represent a cavity embedded in a dielectric. Fig. 1 details this model [17].

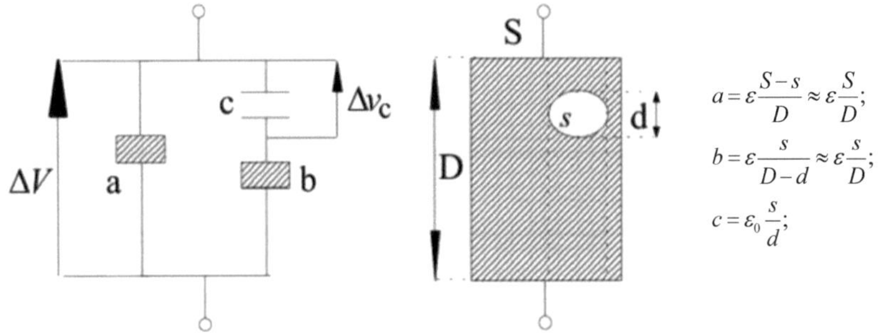

Figure 1. Three capacity of the model of Whitehead.

Fig. 2 presents the V_c voltage applying on the c capacitor. It should be noticed that V_c is a fraction of the total voltage V. When V_c exceeds the so called inception voltage V_{ci}, the discharge starts. After the discharge, V_c reaches the discharge extinction voltage value, V_{ce}. The global behaviour of the system is affected and the PD activity can be monitored by employing current and voltage measures.

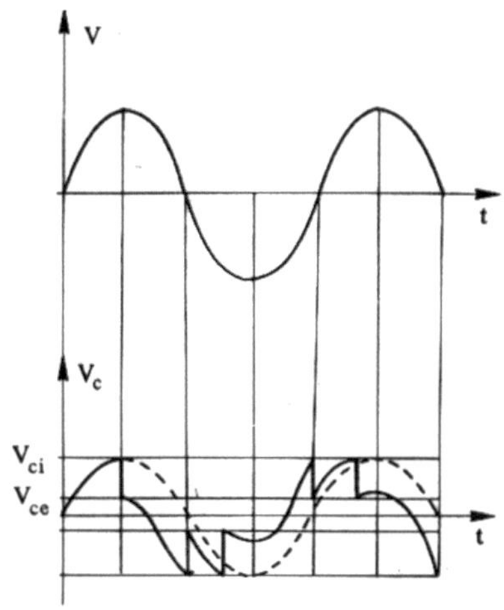

Figure 2. *V* voltage applied to the dielectric, V_c voltage across the cavity, V_{ci} inception voltage of the discharge, V_{ce} discharge extinction voltage.

A field approach model that best represents the phenomenon is conceived by Pedersen [18]. Since the charge involved by the discharge inside the cavity, named true charge, cannot be measured, an auxiliary current, named apparent charge, is measured outside of the specimen. The phenomenon is then traced to two successive stages:

- the first defines the external current measured with the induced charge to the electrodes.
- the second one correlates the induced charge to the electrodes with the charge inside the cavity.

2.2. Partial discharge characterization

A partial discharge is a stochastic phenomenon that involves many chemical and physical aspects of materials proprieties. The PD detection over time has become an indispensable method for the evaluation of insulation systems in the realization phase of new devices but also for evaluate reliability in work. The discharge manifests itself through charge displacement, acoustic waves, lights and electromagnetic field. The wired detection of charge displacement was the most frequently measurement method chosen before the introduction of antenna probes. The basic discharge detection circuit, in accordance with the International Electrotechnical Commission (IEC) Standard, is shown in fig. 3 [19]. The coupling capacitor provides a low impedance path for high-frequency discharge currents which cause voltage pulses over the measuring impedance. The pulses are then amplified and stored in a computer for further processing and displayed in different patterns as 3D patterns (phase-amplitude-number of discharge) and 2D patterns (amplitude-number, phase-number, phase-amplitude).

In the same IEC Standard the main definitions related to PD are defined as follows:

Partial discharge (PD): Localized electrical discharge that only partially bridges the insulation between conductors and which can or cannot occur adjacent to a conductor.

Partial discharge pulse (PD pulse): Current or voltage pulse that results from a partial discharge occurring within the object under test. The pulse is measured using suitable detector circuits, which have been introduced into the test circuit for the purpose of the test.

Apparent charge q: Of a PD pulse is that charge which, if injected within a very short time between the terminals of the test object in a specified test circuit, would give the same reading on the measuring instrument ad the PD current pulse itself. The apparent charge is usually expressed in picocoulombs (pC).

Pulse repetition rate n: Ratio between the total number of PD Pulses recorded in a selected time interval and the duration of this time interval.

Phase angle ϕ_i and time t_i of occurrence of a PD pulse: is $\phi_i=360(t_i/T)$ where t_i is the time measured between the preceding positive going transition of the test voltage through zero and the partial discharge pulse and T is the period of the test *voltage.*

The pulses generated by PDs are detected as second order pulses; in Fig.4 the pulse was recorded by means of an oscilloscope and shows a signal proportional to the discharge current. It was detected by means of a 50 Ω resistance in series to the circuit where the defect was present. This kind of pulses are very fast (~50ns) and have a small amplitude (~5mV). As we will see in the following, not only the amplitude of the pulse but also its duration is a fundamental parameter. The most important recent studies that have allowed PD measurements to become a reliable diagnostic system are based on the shape of the pulses, and in particular on the ratio between duration and frequency content. The acquisition techniques (bandwidth of the recording system) and the kind of sensor adopted (HFCT, capacitive couplers) become fundamental components because they affect the shape of the impulses.

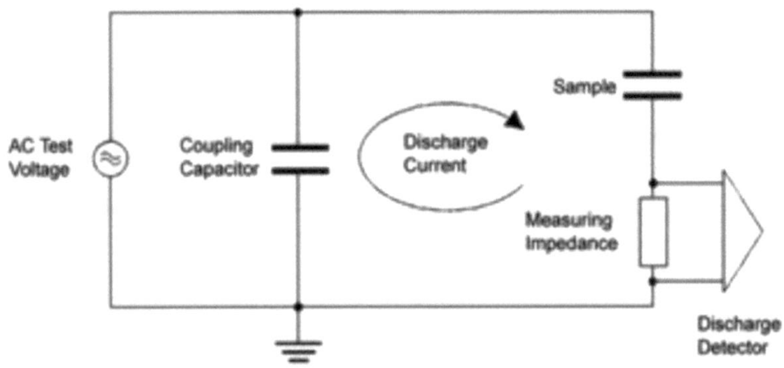

Figure 3. Discharge detection circuit.

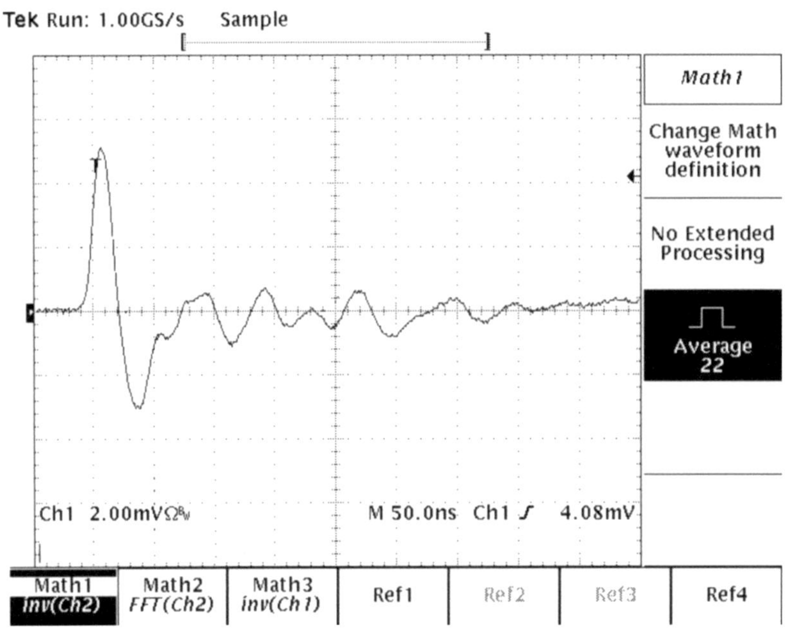

Figure 4. Generic second order PD Pulse.

In the PD detection theory, three families of PD phenomena have been taken into account: namely internal, surface and corona. Internal discharges occur inside a cavity embedded in a dielectric. These are cavities within resin components, extruded plastic cables, joint, stator insulation and external particles such as dust or textile fibres embedded in insulation. Surface discharge may occur along dielectric surfaces with a high tangential field stress. Typical examples of high voltage components that present surface discharges are bushing, twisted pairs, ends of cables, the overhang of generator windings. Corona discharges occur in a needle plane configuration at sharp points at high voltage or ground potential [20].

http://dx.doi.org/10.5772/58840

The pulse shape of singular discharge phenomena present differences in term of time duration and frequency content. These differences are exhibited in Fig. 5 where an example of pulse shape and frequency spectrum for each discharge family is showed.

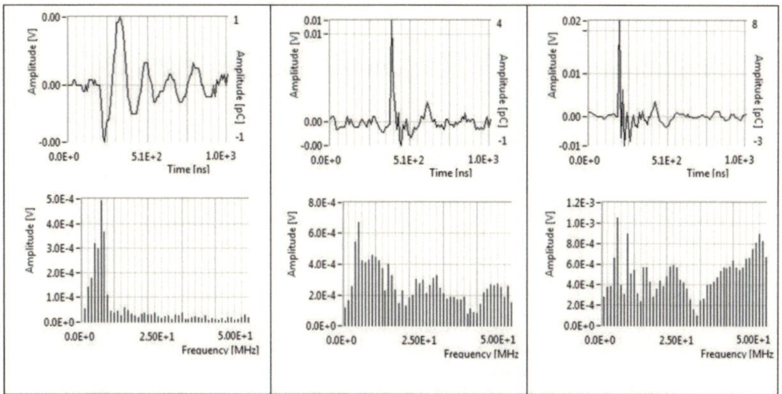

"Slow" pulses	"Middle-fast" pulses	"Fast" pulses
Duration >1ms	Duration <1ms	Duration <100 ns
Spectrum <20 Mhz	Spectrum <40 Mhz	Spectrum <80 Mhz
Typically surface discharges	**Tipically corona discharges**	**Tipically internal discharges**

Figure 5. Time duration and frequency spectrum of surface, corona and internal discharges.

Moreover, the acquisition of the pulse repetition on a time interval ΔT is displayed in a 3D histogram generally known as PD Pattern (phase-amplitude-number of discharge) and 2D PD patterns (amplitude-number, phase-number, phase-amplitude,) as displayed in Fig.6. Furthermore, the pattern allows the understanding of the kind of the defect.We obtain different patterns for each discharge family as shown in Fig.7, 8 and 9.

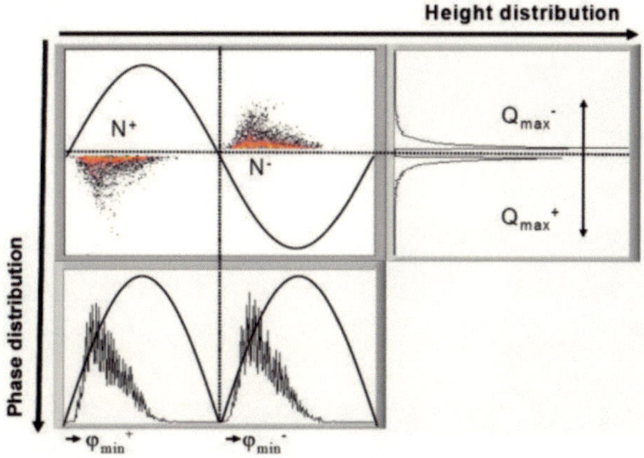

Figure 6. PD Pattern with 2D amplitude-number (right) and phase-number (below). In the 3D PD Pattern the colour scale represents the intensity of the discharges.

Fig.7 shows a typical internal void rabbit-like PD pattern. Internal discharge are characterized by a quite symmetry for positive and negative discharges due to electron emission from the same materials (dielectric). In particular the showed pattern is referred to a spherical air void embedded in a dielectric subjected to a uniform electric field.

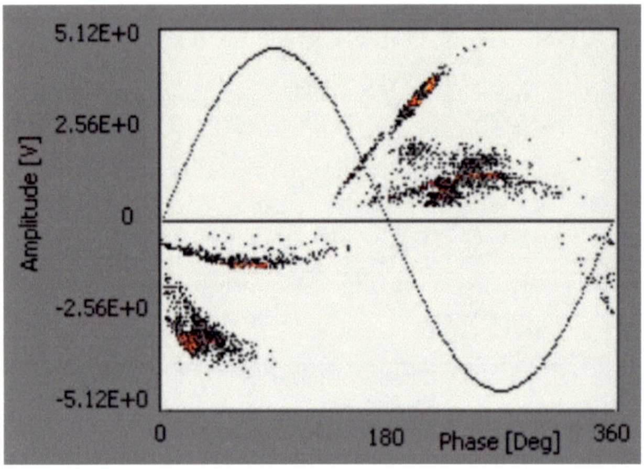

Figure 7. PD Pattern of internal discharges.

Fig.8 shows a typical surface PD pattern, where a substantial dissymmetry between positive and negative discharges can be noted. The inception phase angle, $\Delta\phi$, defined as the phase interval between the first and the last discharge (positive or negative) is smaller than the one observed for internal discharges.

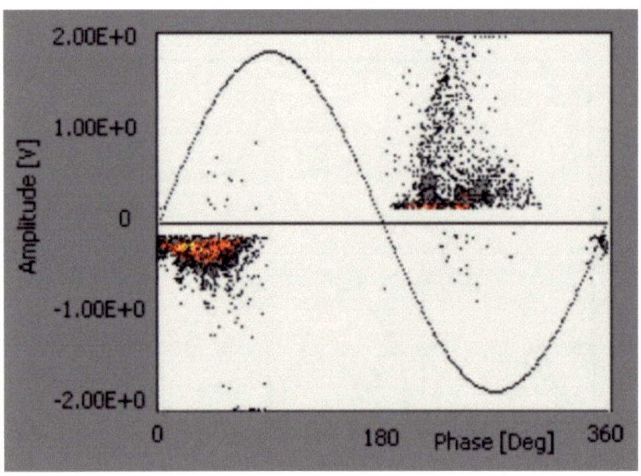

Figure 8. PD Pattern of surface discharges.

Fig.9 shows a typical corona PD pattern. Two characteristic differentiates it from the previous discharges. The first is the presence of discharge only in half period of AC source and the second is that the discharge occurs in correspondence of the voltage pick of AC source, in the other cases, differently, the discharges occur in presence of the maximum gradient of the voltage.

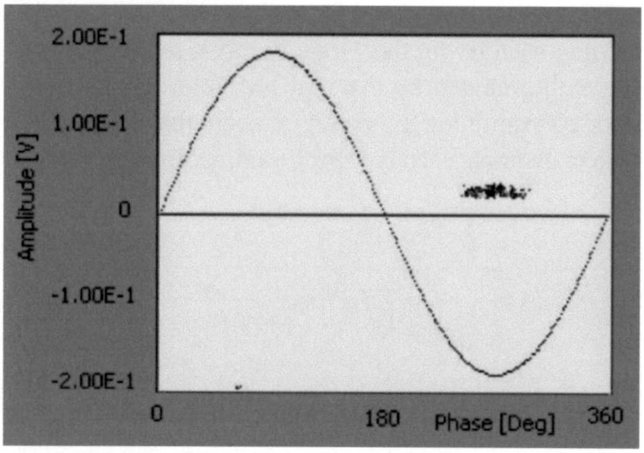

Figure 9. PD Pattern of corona discharges.

In many real cases PD patterns present very different shapes and sometimes with more than only one defect at the same time. For this reason pattern recognition techniques for the discrimination and classification of discharges represent a fundamental instrument for the reliability's evaluation of high voltage electrical apparatus. Many different methods have been proposed in literature to automate the identification of defects through PD pattern recognition. They differ either in presentation of PD data (PD pulse shape data, statistical parameters of these data and pulse density distribution) or in the classification methods (statistical algorithms or neural network (NN)) [20-22]. Recently the use of wavelet analysis and advanced models has provided interesting results in PD denoising processes, pattern recognition and classification of multi-source PD patterns [23,26].

One limit of the classic PD measurement s is due to the wired circuit, signals are dumped or affected by the particular circuit path encountered. In order to overcome these limits wireless measurement have been proposed.

3. Method of moments

The Method of Moments (MoM) in two dimensions has been successfully applied to describe the behaviour of three-dimensional objects that have a highly shielding external structure, for which the hypothesis of perfect electric conductors can be used [11]. Other applications of MoM involving three-dimensional objects composed of voxels or thin wire elements can be found in [12]. We note S the surface of a metal object considered a perfect conductor (PEC) and n the normal to surface S. By considering an Electromagnetic (EM) stress, E_i denotes the incident electric field, which can be found in the absence of the object having S surface. The presence of an EM field generates surface current J_S on S. The aim of the EM analysis is to determine the current in S. Since no analytical solution can be found for complex geometries only a numerical one can be obtained approximating the current J_S on S into J_i currents on $S^N = \cup_1^N S_i$, (see Fig. 10). Triangular elementary surfaces are used to describe the whole S^N, by introducing edge limiting each triangular faces. Vertexes are the points shared by different triangles. Each triangle, illuminated by external EM field, has its own current, which can conceptually be studied by examining the average current that passes through the three edges. The aim of the MoM is to evaluate each current J_i passing through an edge.

The current on the whole S can be approximated by:

$$\vec{J}_S = \sum_{i=1}^{N} J_i \vec{f}_i(\vec{r}), \tag{1}$$

where N is the number of internal sides, and $f_i(\vec{r})$ are a suitable functions, called basis functions.

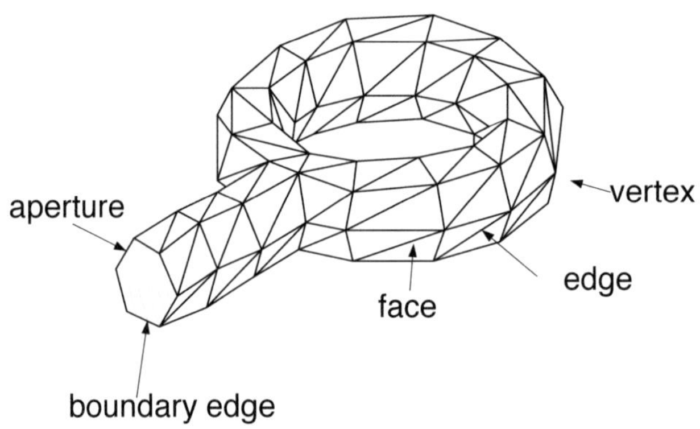

Figure 10. Loop antenna described with triangular surfaces

The basis functions to be used must meet two basic requirements. Indeed, they have to:

- be suitable to be used with an electromagnetic equation.
- Allow modeling with flat triangles. Further useful developments can be found in [27].

Integro-differential equation can be derived by imposing equality between the incident field and scattered field on the boundary condition on S,

$$\hat{n} \times (E_i + E_s) = 0 \tag{2}$$

getting:

$$-E_{inc}^{tan} = -(j\omega\vec{A} - \nabla\varphi)^{tan}, \tag{3}$$

where $\vec{A}$ and φ are the vector and the scalar potential respectively:

$$\vec{A}(\vec{r}) = \frac{\mu}{4\pi}\int_S \vec{J}_S(\vec{r})\frac{e^{-jkR}}{R}dS' \tag{4}$$

$$\varphi(\vec{r}) = \frac{1}{4\pi\varepsilon}\int_S \sigma(\vec{r})\frac{e^{-jkR}}{R}dS', \tag{5}$$

with $k = \omega\sqrt{\varepsilon\mu}$.

Such potentials are depending on the local surface current J_S and on the local charge density σ, which can be related to J_S by taking into account the following:

$$\nabla \bullet J_S = -j\omega\sigma \tag{6}$$

The set of equations described here takes the name of Electric Field Integral Equation (EFIE).

Equations 4 and 5 have no analytical solution for a complex S, so an approximation similar to the one used in equation 1 is required, considering only the unknowns J_i set in equation 1.

By considering an edge i shared by two triangles called T_i^+ and T_i^-, it is possible to describe the current in each point M of the triangles, by using the basis functions(see Fig. 11a).

Points (M) belonging to T_i^+ can be detected by means of the distance $\vec{r}$ from an absolute center **O** or through ρ_i^+distance from the free vertex of T_i^+. Similar remarks can be applied for T_i^-. The choice of the+and-polarities is due to the implementation of a surface current density J_i, which crosses the i side, leaving T_i^+ for T_i^-. In Fig.11 the behaviour of the current in M is associated to the centroids C+ and C-.

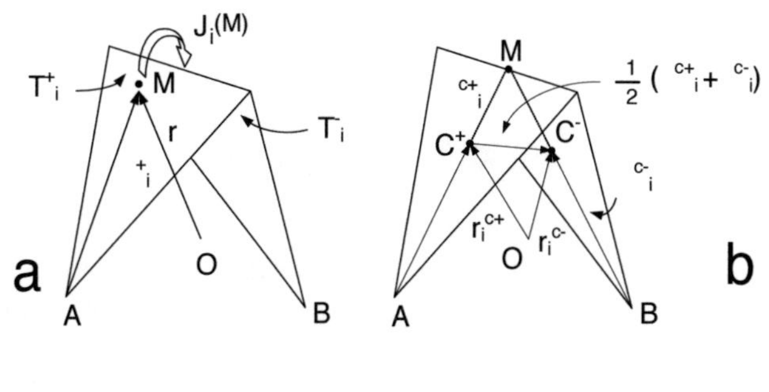

Figure 11. Triangles sharing an internal edge. Focus on the current through the edge (a), focus on centroids.

The purpose of the basis function is to simplify the use of equations 4 and 5, projected in the numerical approximation. The integrals on *S* are solved for each triangle by taking into account three currents leaving the three edges. The basis function vector is defined by:

$$\vec{f}_i(\vec{r}) = \begin{cases} \frac{l_i}{2A_i^+}\vec{\rho}_i & r in T_i^+ \\ \frac{l_i}{2A_i^-}\vec{\rho}_i & r in T_i^- \\ 0 & elsewhere \end{cases}, \tag{7}$$

Where l_i is the length of the edge taken into consideration, A_i^+ and A_i^- are the areas of the triangles. Also the following two features are required:

$$\nabla \bullet \vec{f}_i(\vec{r}) = \begin{cases} \frac{l_i}{A_i^+} & r in T_i^+ \\ -\frac{l_i}{A_i^-} & r in T_i^- \\ 0 & elsewhere \end{cases}, \tag{8}$$

$\vec{f}_i(\vec{r}) = \vec{0}$ on the boundary of *S*.

It is not possible to solve equation 3 on the whole S, but it can be solved on $S^N = \cup_1^N S_i$, by employing the numerical projection on a N-dimension system. Similarly to the Fourier's sampling a particular function has to be used to project in the N-dimension space, instead of the delta of dirac function here are used the basis functions:

$$\langle \vec{X}, \vec{f} \rangle = \int_S \vec{X} \bullet \vec{f} \, dS, \tag{9}$$

which in the *N-space* becomes:

$$\langle \vec{X}, \vec{f} \rangle \cong \sum_{i=1}^{N} \vec{X}_i \bullet \vec{f}_i A_i, \tag{10}$$

with $\vec{X}$ a vector function. In such a way equation 3 becomes:

$$\left\langle E_{inc}^{tan}, \vec{f} \right\rangle = \langle j\omega \vec{A}, \vec{f} \rangle^{tan} + \langle \nabla \varphi, \vec{f} \rangle^{tan}. \tag{11}$$

In equation 11 the gradient is applied on the scalar potential, but considering the following identity (similar to the well known Stokes theorem), it is possible to apply it on the basis function:

$$\langle \nabla \varphi, \vec{f} \rangle = \int_S \nabla \varphi \bullet \vec{f} \, dS = \int_S \nabla \bullet (\varphi \vec{f}) dS - \int_S \varphi \nabla \bullet \vec{f} \, dS = \int_{\partial S} \varphi \vec{f} \, dS - \int_S \varphi \nabla \bullet \vec{f} \, dS = -\langle \varphi, \nabla \bullet \vec{f} \rangle, \tag{12}$$

http://dx.doi.org/10.5772/58840

with $\vec{f}$ equals to zero on the boundary of both S and ∂S. Finally, by considering a triangle p it is possible to write equation 11 as follows:

$$\left\langle E_{inc}^{p,tan}, \vec{f}_p \right\rangle = \left\langle j\omega \frac{\mu}{4\pi} \int_S \vec{J}_S(\vec{r}) \frac{e^{-jkR}}{R} dS', \vec{f}_p \right\rangle^{tan} - \left\langle \frac{1}{4\pi\varepsilon} \int_S \sigma(\vec{r}) \frac{e^{-jkR}}{R} dS', \nabla \bullet \vec{f}_p \right\rangle^{tan}. \tag{13}$$

To study the effects of the triangle p on the current due to the current of a triangle q(seeFig. 12.a), all the edge currents have to be considered. In Fig.12.b, the interaction between currents on edge of length l_m of p and edge of length l_i of q,is shown.

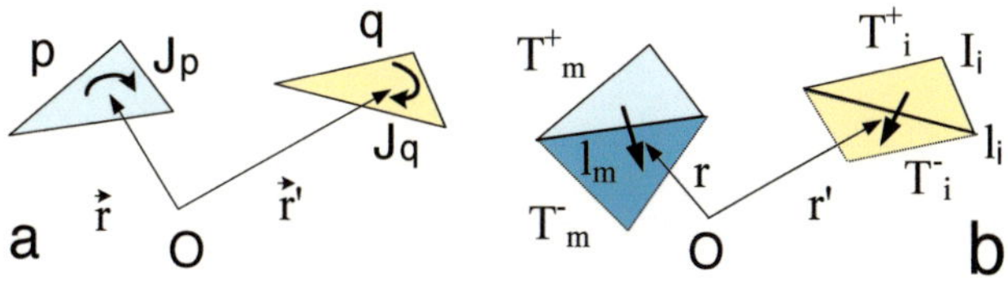

Figure 12. Interactions between currents of different triangles

In equation 13 the inner product has to be considered twice: the formal inner product with $\vec{f}_p$ and the one inside surface integral $\int_S dS'$. By considering initially the first inner product, for edge m, shared by T_m^+ and T_m^-, and also considering value on the centroid as a constant value for all triangle, we can write:

$$\begin{gathered}\left\langle E_{inc}^m, \vec{f}_m \right\rangle = \int_{T_m^+ + T_m^-} E_{inc}^m \vec{f}_m dS' = E_{inc}^{m,c+} \int_{T_m^+} \vec{f}_m dS' + E_{inc}^{m,c-} \int_{T_m^-} \vec{f}_m dS' = \\ E_{inc}^{m,c+} \left(\frac{l_m}{2A_m^+} \rho_m^{c+} \right) A_m^+ + E_{inc}^{m,c-} \left(\frac{l_m}{2A_m^-} \rho_m^{c-} \right) A_m^- = \frac{1}{2} E_{inc}^{m,c+} l_m \rho_m^{c+} + \frac{1}{2} E_{inc}^{m,c-} l_m \rho_m^{c-},\end{gathered} \tag{14}$$

$$\begin{gathered}\left\langle j\omega \vec{A}(\vec{r}_m^{c+}), \vec{f}_m \right\rangle^{tan} + \left\langle j\omega \vec{A}(\vec{r}_m^{c-}), \vec{f}_m \right\rangle^{tan} = \int_{T_m^+} j\omega \vec{A}(\vec{r}_m^{c+}) \vec{f}_m dS' + \int_{T_m^-} j\omega \vec{A}(\vec{r}_m^{c-}) \vec{f}_m dS' = \\ \frac{1}{2} j\omega \vec{A}(\vec{r}_m^{c+}) l_m \rho_m^{c+} + \frac{1}{2} j\omega \vec{A}(\vec{r}_m^{c-}) l_m \rho_m^{c-}\end{gathered} \tag{15}$$

$$\begin{gathered}\left\langle \varphi(\vec{r}_m^{c+}), \nabla \bullet \vec{f}_m \right\rangle^{tan} + \left\langle \varphi(\vec{r}_m^{c-}), \nabla \bullet \vec{f}_m \right\rangle^{tan} = \\ \int_{T_m^+} \varphi(\vec{r}_m^{c+}), \nabla \bullet \vec{f}_m dS' + \int_{T_m^-} \varphi(\vec{r}_m^{c-}), \nabla \bullet \vec{f}_m dS' = \varphi(\vec{r}_m^{c+}) l_m - \varphi(\vec{r}_m^{c-}) l_m.\end{gathered} \tag{16}$$

The second inner product is inside the vector and scalar potentials, that are due to the current on the edge i:

$$\vec{A}(\vec{r}_m^{c\pm}) = \frac{\mu}{4\pi} \int_{T_i^{\pm}} J_i \vec{f}_i \frac{e^{-jkR_m^{\pm}}}{R_m^{\pm}} dS' = \frac{\mu}{8\pi} J_i l_i \rho_i^{c\pm} \frac{e^{-jkR_m^{\pm}}}{R_m^{\pm}}, \tag{17}$$

where $R_m^{\pm}$ is the distance of the centroids of triangles T_m^+ and T_m^- from the ones of T_i^+ and T_i^-.

In similar way:

$$\varphi\left(\vec{r}_m^{\,c\pm}\right) = -\frac{1}{j\omega 4\pi}\int_{T_i^{\pm}} J_i \nabla \bullet \vec{f}_i \frac{e^{-jkR_m^{\pm}}}{R_m^{\pm}} dS' = \mp \frac{1}{j\omega 4\pi} J_i l_i \frac{e^{-jkR_m^{\pm}}}{R_m^{\pm}}. \tag{18}$$

The double inner product gives a symmetry to the system, the actions of the current *i* on *m* and of *m* on *i* are similar. The total effect on *m* is due to an extended summation of index *i*. In such a way the following equation system holds:

$$ZJ = V, \tag{19}$$

with $Z = \left[Z_{m,i}\right]$,

$$Z_{m,i} = \frac{1}{2} j\omega \frac{\mu}{8\pi} \frac{e^{-jkR_m^+}}{R_m^+} l_i \rho_i^{c+} l_m \rho_m^{c+} + \frac{1}{2} j\omega \frac{\mu}{8\pi} \frac{e^{-jkR_m^-}}{R_m^-} l_i \rho_i^{c-} l_m \rho_m^{c-} + \frac{1}{j\omega 4\pi} \frac{e^{-jkR_m^+}}{R_m^+} l_m l_i - \frac{1}{j\omega 4\pi} \frac{e^{-jkR_m^-}}{R_m^-} l_m l_i, \tag{20}$$

$$J = \begin{bmatrix} J_1 \\ \vdots \\ J_N \end{bmatrix}, \tag{21}$$

$$V = \begin{bmatrix} \frac{1}{2} E_{inc}^{1,c+} l_1 \rho_1^{c+} + \frac{1}{2} E_{inc}^{1,c-} l_1 \rho_1^{c-} \\ \vdots \\ \frac{1}{2} E_{inc}^{N,c+} l_N \rho_N^{c+} + \frac{1}{2} E_{inc}^{N,c-} l_N \rho_N^{c-} \end{bmatrix} \tag{22}$$

The vector *V* contains in each row the effect of the incident electric field due to an external source; if the source is an internal voltage generator, all rows are zero except the one in which is enforced the electric field due to the generator, approximated by the imposed voltage divided by the distance between centroids.

Once defined equation 19, the unknown currents *J* can be evaluated on the whole surface of the object (see Fig.13.a). Each current starts and ends on a centroid. A dipole moment is definable for each current J_i crossing an internal edge of length l_i(see Fig.11.a):

$$\vec{m}_i = \int_{T_i^+ + T_i^-} J_i \vec{f}_i dS' = \frac{1}{2} l_i J_i \left(\vec{\rho}_i^{\,+} + \vec{\rho}_i^{\,-}\right) = l_i J_i \left(\vec{r}_i^{\,-} - \vec{r}_i^{\,+}\right) \tag{23}$$

The radiated magnetic and electric field, due the dipole of current J_i, in a generic point P (Fig. 12.b) located at a distance $\vec{r}$, are given by:

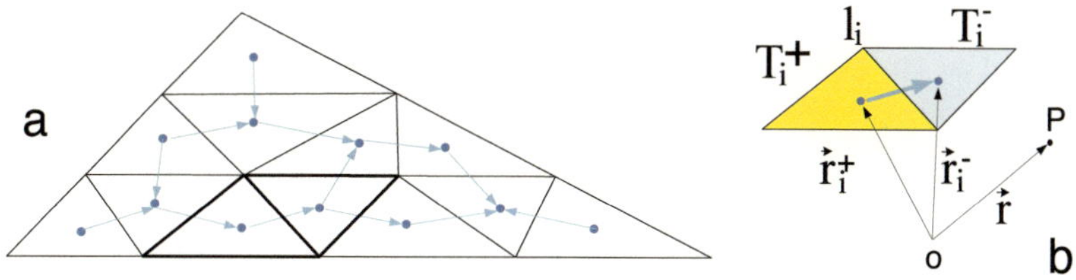

Figure 13. A triangular antenna divided into triangular subparts, in which internal current are shown (a). Particular of the dipole model for the current crossing edge *i* (b).

$$\vec{H}_i(\vec{r}) = \frac{jk}{4\pi}\left(\vec{m}_i \times \vec{r}\right)\frac{1}{r^2}\left(1 + \frac{1}{jkr}\right)e^{-jkr}, \tag{24}$$

$$\vec{E}_i(\vec{r}) = \frac{\eta}{4\pi}\left(\left(\frac{(\vec{r} \bullet \vec{m}_i)\vec{r}}{r^2} - \vec{m}_i\right)\left(\frac{jk}{r} + \frac{1}{r^2} + \frac{1}{jkr^3}\right) + 2\frac{(\vec{r} \bullet \vec{m}_i)\vec{r}}{r^4}\left(1 + \frac{1}{jkr}\right)\right)e^{-jkr}. \tag{25}$$

4. Antenna modeling with method of moments

The aim of this study is to create a model to describe the behaviour of PD signal probe. Antennas can be easily designed with commercial software like Matlab® in which suitable tools also create the Delaunay triangulation.

The simplest antenna to be designed is the monopole antenna, shown in Fig.14. In order to create the 20 cm long monopole two rectangles have been created, one thin in which only a line of triangle is needed, another bulky.By considering the possible interactions between the two triangles, problems arise. There is an edge shared between three triangles (red, yellow, blue). The current of this edge will be divided into 2 sub-currents and another row and unknown is set in equation (6).

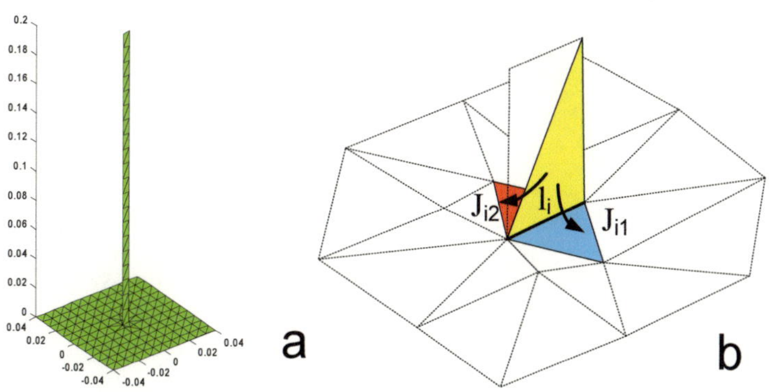

Figure 14. Discrete surface MoM model of the monopole antenna.

Similarly all non-planar antennas are treated. In Fig. 15 the triangular antenna and the spherical one are shown. The triangular antennas simulates a planar 20 cm equilateral triangle. The sphere has 6 cm diameter.

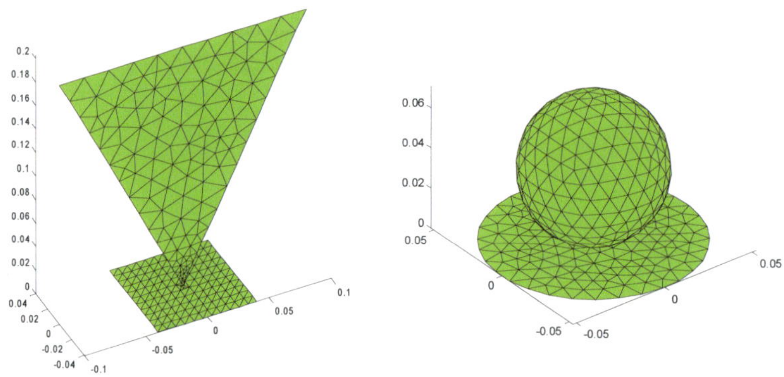

Figure 15. Discrete surface MoM model of the triangular and spherical antennas.

In Fig.16 an electronic power device is sketched: two metal plates, normally at different voltage, during a partial discharge are metallically linked by the discharge path.

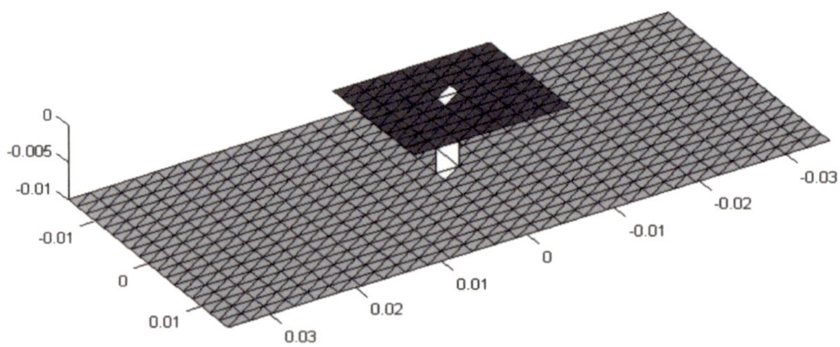

Figure 16. Discrete surface MoM model of the emitting source.

The numerical model allows to feed all the antennas in two different ways. The first one uses a voltage generator, set between the antenna (upper object) and ground plate (bottom metallic plate). The other way takes into account the illuminating external field on the surface of the whole object. In order to simulate the transmission of fault EM signal and its receiving, the power device is fed by generator and the other antennas are used as probes, so the provided voltage on the linking edge is studied.

http://dx.doi.org/10.5772/58840

As previously explained the source of the EM field illuminating the probes is given by the discharge current flowing through the discharge path, which is due to the ionized air gap inside of insulation, and embraces previously damaged materials and metallic surfaces. A full characterization of the exact discharge channel is beyond the scope of this work and also for sake of simplicity the discharge path is simulated with a metallic wire in which the discharge current flows unhindered. Then neglecting particular behaviour to the geometry of the discharge channel, scrupulous attention is paid to the waveform of the current, the harmonic content, which will characterize the fidelity of the receiving probes, is studied.

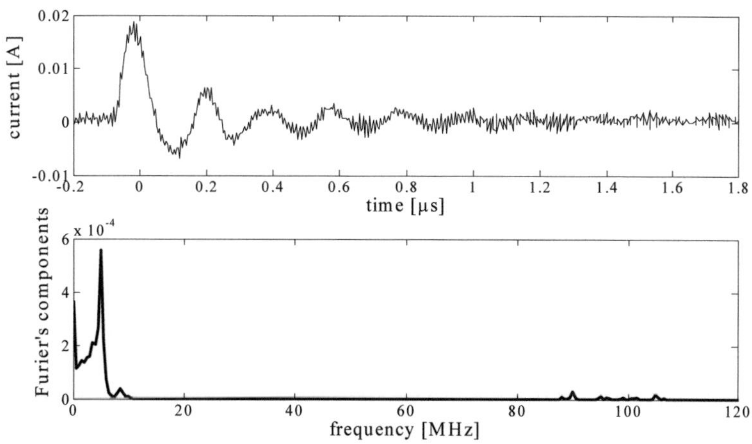

Figure 17. Discharge current and Fourier spectrum. Noise is on 90-110 MHz range.

In Fig.17, a PD signal is represented, which is characterized by an underdamped oscillation. This latter propagates until the energy balance is once again stable, and the charges back to be imprisoned in the dielectric. In order to study the transient PD signal Discrete Fast Fourier Transform (DFFT) has to be used. The PD signals may have a faster or slower evolving, but roughly the spectrum of interest is contained in a [0-200MHz] window. This aspect allows characterizing the type of receiving probes. Unfortunately, in the environment in which PD signals are generated, other signals, that should affect in a limited way the diagnostic tool, are present. PD signal of Fig.17intentionally contains a part of noise that will interact with the probes receivers.

Once defined the behaviour of the PD current, it is possible to characterize the behaviour of the source system, focusing at 20 cm from the discharge, chosen as a compromise between the possibility of bringing a wireless sensor to the faulty part and the need to maintain a safe distance or in other cases distance imposed by the presence of screens, shields and spacers.The PD discharge path has a normal direction with respect to the ground plate and an unitary current source is applied within the range of [0-1000MHz]. The simulated z-component of the electric field at a distance of 20 cm, is shown in Fig.18. The electric field profile is growing with the frequency, similar to the one of a dipole with its dimension equal to the discharge path.

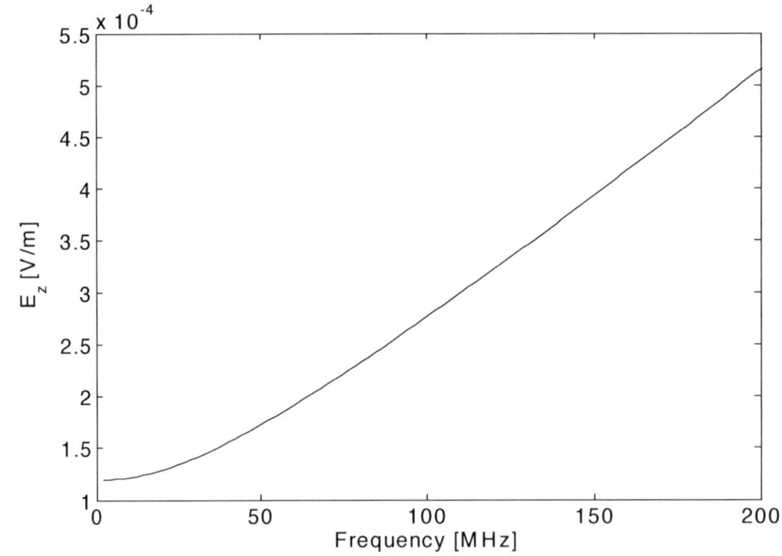

Figure 18. E_z field generated by the discharge at a horizontal distance of 20 cm. Only the 0-200 MHz interval is shown.

In order to compare the different antennas behaviour a particular transfer function is considered:

$$H = \frac{V_{received}}{E_{incident}} \left[m \right] \tag{26}$$

where $V_{received}(\omega)$ is the voltage response of antennas, illuminated by the incident electromagnetic field.

The incident EM field due to the PD current, which is function of the signal frequency content, has to be computed. The received voltages can be computed by simulating the response of the antenna illuminated by the incident EM field. The voltage response is due to both source and antenna impedance behaviour. The three transfer functions corresponding to the three antenna typologies are plotted in Fig. 19considering the above mentioned frequency range.

As it can be seen, the monopole antenna provides the highest gain but its profile is characterized by a resonant peak centered at about 450 MHz. The triangular antenna shows a resonant peak centered in about 400 MHz but with lower amplitude with respect to the former. On the contrary, the spherical antenna shows a flat bandwidth with the lowest amplitude among the three.

http://dx.doi.org/10.5772/58840

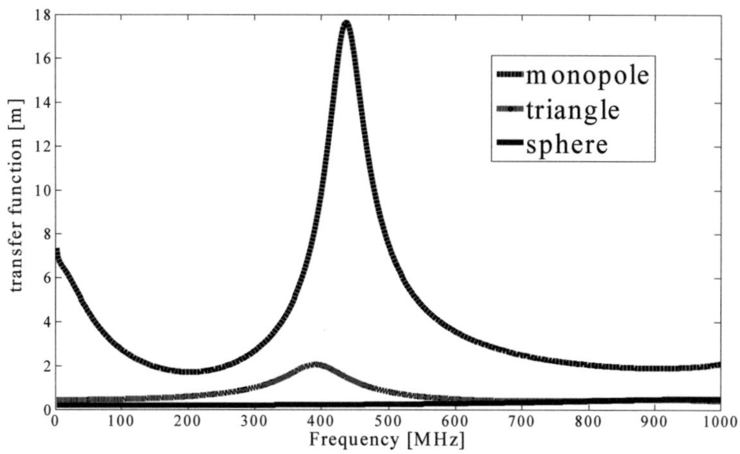

Figure 19. Transfer function of the three antennas.

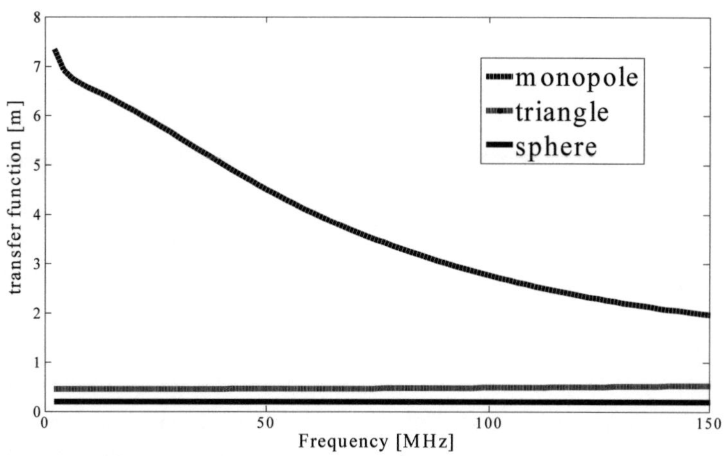

Figure 20. Transfer function of the three antennas for the range of 0-150MHz.

In Fig.20, the different transfer functions are plotted in the frequency range going from 0 to 200 MHz. The spherical and triangular antennas show an almost flat behaviour while the monopole is characterized by a significant decrease in amplitude. It is easily to argue that these different profiles have a different impact on the shape of the recorded PD signals. The selection of a specific profile depends on the application, e.g. PD measurements in dc, ac and rectangular (adjustable speed drive applications) voltage supply. A procedure able to evaluate the signal distortion of the different antennas has been developed to select the antenna able to fit the separation and identification requirements.

5. Transmission of Signals

In order to test the different antennas, bursts of damped pulses were simulated with superimposed high frequency noise in the frequency range of 0-200MHz. The performances of the three antenna configurations were evaluated considering the simulated received signals. An example of the normalized voltage response of the three antennas is plotted in Figure 21. According to Fig. 19, it was found an amplitude reduction of-10 dB and-14 dB for the triangular and spherical antennas with respect to the monopole probe.

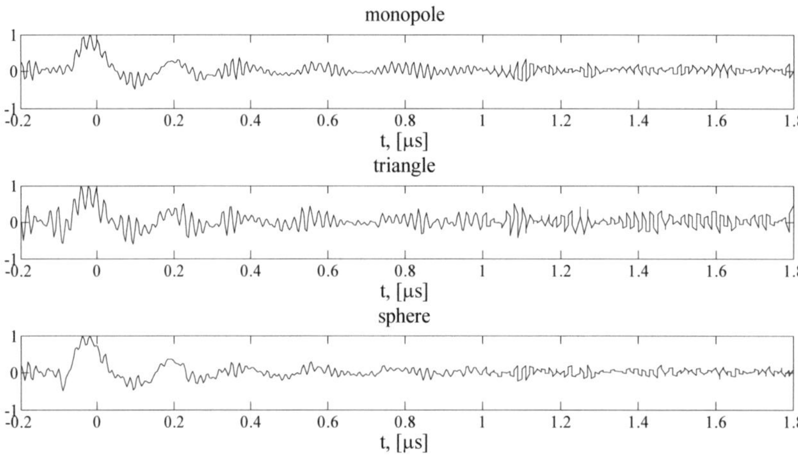

Figure 21. Time profiles of the received voltage.

Moreover, the high frequency noise has the highest influence on the pulse signal received by the triangular antenna since its transfer function reduces the amplitude of the pulse signals and amplifies the superimposed noise. The opposite considering the monopole configuration, while the flat bandwidth of the spherical antenna reduces both the amplitude and the superimposed noise of the received pulse signal. A dedicated procedure has been developed in order to validate this empirical evaluation.

The proposed validation procedure is based on the classification of the autocorrelation function (ACF) of the PD signals, via the K-Means classification algorithm [10]. The ACF has been proposed for its ability to synthesize both the time and frequency domain features as well as to be less affected by superimposed high-frequency noise, random truncation of the pulse-tail (frequency leakage) and different trigger activation. The ACF is evaluated by taking into account the following expression:

$$R_S = \int_{-\infty}^{\infty} s(t)s(t+\tau)dt \tag{27}$$

where *s(t)* represents the time behaviour of the simulated PD signal in discrete time domain. In order to compare the shape of signals with different energy, the normalized ACF in discrete time domain has been here adopted:

$$R_S[k] = \frac{\sum_{i=1}^{N} s(i)s(i+k)}{N}\frac{1}{E_S}, \tag{28}$$

Where *s(i)* is the signal digital sample, *N* is the number of digital samples, E_S is the energy of the signal:

$$E_S = \frac{\sum_{i}^{N} s(i)^2}{N} \tag{29}$$

Fig.s22 and 23 show the NACF of the signals reported in Fig. 17 and 21, respectively.

The Pearson correlation index, ϱ, is used to compare the different NACF with the emitted one. The covariance between two NACFs can be evaluated by the following expression:

$$cov(R_{S1}, R_{S2}) = \frac{\sum_{k=1}^{N}\left[R_{S1}(k) - \bar{R}_{S1}\right] \bullet \left[R_{S2}(k) - \bar{R}_{S2}\right]}{N-1}, \tag{30}$$

where $\bar{R}_{S1}$ and $\bar{R}_{S2}$ are the mean value of NACFs.

The Pearson correlation index is:

$$\rho = \frac{cov(R_{S1}, R_{S2})}{\sqrt{var(R_{S1})} \bullet \sqrt{var(R_{S2})}}, \tag{31}$$

where $var(R_{S1})$ and $var(R_{S2})$ are the variance of the NACFs.

The distance in the NACF metric can be assumed as:

$$d_\rho = \sqrt{2(1-\rho)} \tag{32}$$

Previous investigations established a $d_\varrho < 0.4$ as a good threshold level in the PD shape comparison. The differences between the transmitted and the received signals using the three different configurations, evaluated considering the distance of equation 32, are reported in Table 1.

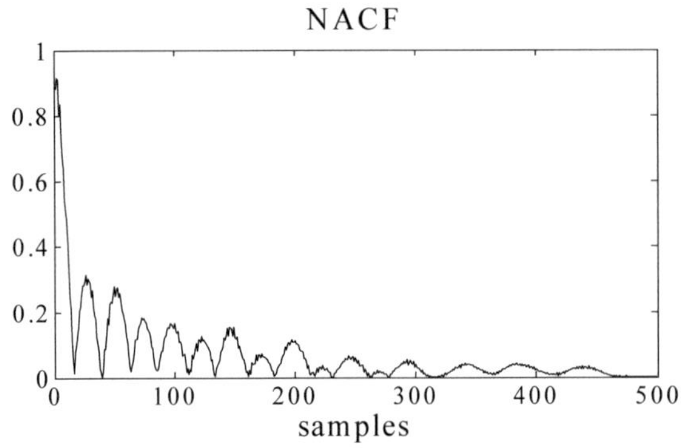

Figure 22. NACF of the PD current of Figure17.

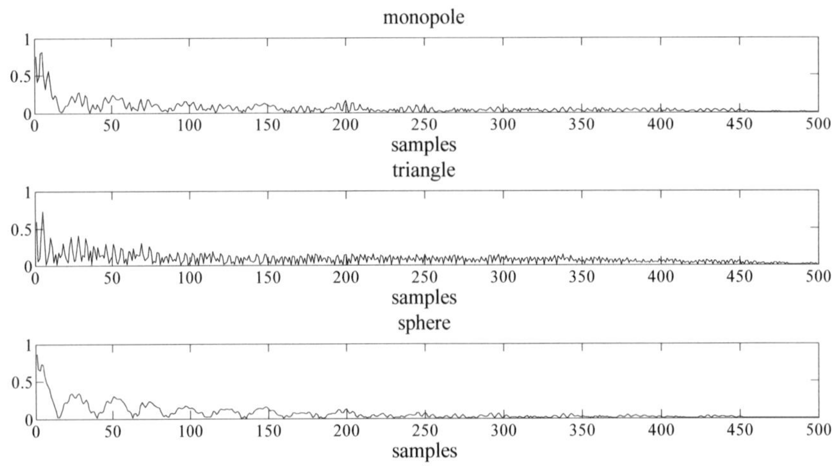

Figure 23. NACFs of the transmitted signals of Figure21.

Distance		
Monopole	**Triangle**	**Sphere**
0.39	0.96	0.29

As it can be seen, the monopole and the spherical antenna preserve the pulse shape in almost similar way while the triangle's received signal is affected by noise and it is groped far from

http://dx.doi.org/10.5772/58840

the PD signal. Due to the different attenuation in the different frequency ranges (Fig. 19), the monopole antenna tends to modify differently the shape of signals generated by the same source. This can induce an over classification in separation procedure that is, the generation of more classes than PD sources. The spherical antenna, due to its flat bandwidth, seems to be the most suitable wireless probe for ac applications.

6. Discussion

The table I shows that the most suitable probes for the identification of partial discharge signals are the monopole probe and the spherical probe. This evaluation was obtained using one of the simplest classification algorithms, certainly other more advanced algorithms may change this rating, and also the use of digital filters in the same implementation may overturn the comparison. The use of filters or the use of more elaborate techniques of signal processing goes beyond the intentions of this chapter. Therefore, the comments will be expressed only in relation to the analogical performance of probes.

6.1. Monopole antenna

The monopole probe has a high gain, superior to the others. This is one of the key aspects in the evaluation of partial discharge phenomena that makes it attractive: the discharges take place within structures often shielded, for which the outgoing signals are weak. The use of too attenuating probe could make the measurement insufficient and the monopole can solve this issue. However, the use of this probe in the industrial environment is not wise since it could involve the absence of higher harmonics. Indeed, their contribution can modify the shape of the signal and even changing the initial instant of the discharge. Such behaviour, in the diagnosis of PD phenomenon, is not acceptable, since the determination of the type of discharge (internal, surface, corona, etc..) is based on the presence of a greater or lesser occurrence of discharges at certain angles of the sine wave function. One way to change the behaviour of the monopole is to change its length, but this length cannot be sufficiently modified to remove the effect of resonance. A simple way to try to mitigate the effect of the resonance is to insert appropriate RLC circuits. UWB antennas are actually filters, which introduce dependences between the transmitter and the receiver frequencies. Therefore such kind of antennas can be treaded. Generally, antennas have linear and passive behaviour, and input impedance can be represented by means of the canonical forms of Foster [28-30].The structure shown in Fig.24 is suitable to represent electric antenna as the electric dipole and monopole.In order to modify the behaviour of the antenna, the electric model of the antenna is required. The numerical implementation of the antenna gives the complex impedance used in the equation 19. The equivalent circuit has to show a similar behaviour for all the harmonic of interest. By following the steps in [28] the value of C_0 is set by taking the value of the capacity presented by the antenna at low frequencies; the value of L_0 is initially set to cancel the imaginary part of impedance at the first resonance frequency. It is required that the first RLC parallel branch has a resonant frequency corresponding to the first maximum of the real part of impedance. By adding more RLC parallel branch the impedance profile can be easily reproduced. Once the circuital model of the antenna is reproduced, new real RLC parallel branch can be used to modify the behaviour of the antenna by mitigating the resonance effect.

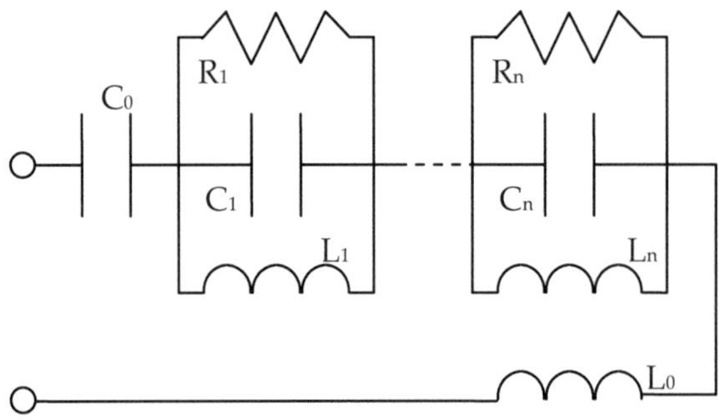

Figure 24. Foster's canonical form for an electrical antenna.

6.2. Triangle antenna

The triangular antenna has a resonance similar to the one of the monopole and shares the same height. The triangular antenna presents an uninteresting behaviour already even used under the best conditions (propagation orthogonal to the surface, absence of high frequencies). Therefore, it is not useful to make changes to improve performances.

6.3. Spherical antenna

The spherical antenna is the true omnidirectional antenna and its performance does not vary by changing its inclination. Then, it presents suitable features to build the PD probes. However, it has a small gain, but this issue can be overcome by using appropriate amplifiers.

7. Conclusion

In this chapter the design of PD receiving antennas has been shown. Three different antennas were simulated by means of the method of moments. Received PD signals have been analyzed by the employment of the classification algorithm, based on the autocorrelation function (ACF) of the signals. The difference in the classification of the received signals has been explained by studying the aspects of the transmitted signals due to the shape of the receiving antennas. The best performances were obtained with the spherical antenna. Indeed, it exhibits an omnidirectional pattern and reliability for reproducing the shape of the partial discharge signal. Thus, this kind of antennas can be suitable to acquire PD pattern to determine the particular type of PD activity.

http://dx.doi.org/10.5772/58840

Author details

Fabio Viola* and Pietro Romano

*Address all correspondence to: fabio.viola@unipa.it

Department of Energy, Information engineering, Mathematical models – DEIM, University of Palermo, Palermo, Italy

References

[1] Yin, X.-C., Ruan, C.-L., Mo, S.-G., Ding, C.-Y., and Chu, J.-H. A compact ultra-wide band microstrip antenna with multiple notches. Progress In Electromagnetics Research, PIER 84, 2008, 321–332.

[2] Lamultree S., and Phongcharoenpanich, C. Bidirectional ultra-wide band antenna using rectangular ring fed by stepped monopole. Progress In Electromagnetics Research, PIER 85, 2008, 227–242.

[3] Akhoondzadeh-Asl, L., Fardis, M., Abolghasemi, A., and Dadashzadeh, G. Frequency and time domain characteristic of a novel notch frequency UWB antenna. Progress In Electromagnetics Research, 2008, PIER 80, 337–348.

[4] Notice of Proposed Rule Making, Revision of Part 15 of the Commission's Rules Regarding Ultra-Wideband Transmission Systems. Federal Communications Commission (FCC), ET Docket 98-153, 2002.

[5] B. A. Fruth and D. W. Gross, Phase Resolving Partial Discharge Pattern Acquisition and Frequency Spectrum Analysis. IEEE 4th Intern. Conf. Propertie, Application of Dielectr. Materials (ICPADM), Brisbane, Australia, paper 6207, 1994.

[6] Baker, P.C. and Judd, M.D. and McArthur, S.D.J.A frequency-based RF partial discharge detector for low-power wireless sensing. IEEE Transactions on Dielectrics and Electrical Insulation2010; 17(1) 133-140.

[7] P. J. Moore, I. E. Portugues, and I. A. Glover, Partial Discharge Investigation of a Power Transformer Using Wireless Wideband Radio-Frequency Measurements.IEEE Trans. On Power Delivery2006; 21(1), 528-530.

[8] A. Cavallini, A. Contin, G. C. Montanari and F. Puletti, Advanced PD Inference in On-Field Measurements. Part.1: Noise Rejection. IEEE Trans. on Dielectrics and Electrical Insulation 2003;10(2) 216-224.

[9] A. Cavallini, M. Conti, A. Contin, G. C. Montanari Advanced PD inference in on-field measurements. Part 2: Identification of defects in solid insulation. IEEE Trans. on Dielectrics and Electrical Insulation 2003; 10(3) 528-538.

[10] A. Contin, S. Pastore, Classification and separation of partial disharge signals by means of their auto-correlation function evaluation. IEEE Trans. On Dielectrics and Electrical Insulation2009; 16(6)1609-1622.

[11] S. M. Rao, D. R. Wilton AND A. W. Glisson, Electromagnetic Scattering by Surfaces of Arbitrary Shape. IEEE Trans. On Antennas snd Propagation 1982; 30(3), 626-629.

[12] Wang, J. H., Generalized moment methods in electromagnetics, Wiley Interscience Publication, 1991.

[13] Makarov, S. N., Antenna and EM Modelling with Matlab, Wiley Interscience Publication, 2002.

[14] G. Ala, R. Candela, F. Viola, Detection of Radiated EM Transients by Exploiting Compact Spherical Antenna Features, Recent Patents on Electrical Engineering 2011, 4, 202-208.

[15] J.C. Anderson. Dielectrics; Chapman & Hall, London 1964.

[16] L. Simoni. Proprietà dielettriche e scarica dei materiali isolanti elettrici; Clueb, Bologna, 1996.

[17] L.A. Dissado, J.C. Fothergill. Electrical Degradation and Breakdown in Polymers, Peregrinus, 1992.

[18] A. Pedersen. Partial Discharges in Voids in Solid Dielectrics. An Alternative Approach. 1987 Annual Report Conference on Electrical Insulation and Dielectric Phenomena, IEEE publication 87 CH2462-0, 1987, 58-64.

[19] IEC Standard 60270. High-Voltage Test Techniques-Partial Discharge Measurements; 2000.

[20] A. Krivda. Recognition of Discharges. Discrimination and classification. Delft University Press, Delft, 1995.

[21] E. Gulski. Computer-aided Recognition of Partial Discharges Using Statistical Tools; Delft University Press, Delft, 1991.

[22] R. Candela, G. Mirelli, R. Schifani PD Recognition by Means of, Statistical and Fractal Parameters and Neural Network. IEEE Transactions on Dielectric and Electrical Insulation 2000; 7(1), 87-94.

[23] X Ma, C. Zhou,I.J. Kemp Interpretation of wavelet analysis and its application in partial discharge detection. IEEE Trans. on Dielectric and Electrical Insulation 2002; 9(3) 446-457.

[24] R. Candela, P. Romano. An improved MSD-based method for PD pattern recognition. 2007 Annual Report Conference on Electrical Insulation and Dielectric Phenomena, October Vancouver BC, Canada October 14-17, 2007, 204-207.

[25] Ala G, Candela R, Romano P, Viola F. Simplified Hybrid PD Model in Voids. 8th IEEE Symposium on Diagnostics for Electrical Machines, Power Electronics &

Drives. Bologna, 5-8 September 2011, 451-455, IEEE, ISBN: 978-1-4244-9303-6, doi: 10.1109/DEMPED.2011.6063662.

[26] Di Silvestre L., Romano P., Viola F. Simplified Hybrid PD model in Voids: Pattern Validation, 4th International Conference on Power Engineering, Energy and Electrical Drives –Powereng, 13-17 may 2013, Istanbul,Turkey.

[27] A. W. Glisson, On the development of numerical techniques for treating arbitrarily-shaped surfaces. Ph.D. dissertation, Univ. Mississippi, 1978.

[28] Foster, R. M. A reactance theorem. Bell Systems Technical Journal, 1924;3(2) 259–267.

[29] S. B. T. Wang, A. M. Niknejad, R. W. Brodersen, Circuit modeling methodology for UWB omnidiretional small antennas. IEEE Journal on Selected Areas in Communications2006; 24(4)871-877.

[30] M. Hamid, R. Hamid, Equivalent Circuit of Dipole Antenna of Arbitrary Length. IEEE transactions on Antennas and Propagation 1997; 45(11) 1695-1696.